The Impact and Legacy of *The Ladies' Diary* (1704–1840): A Woman's Declaration

AMS/MAA | SPECTRUM

VOL **101**

The Impact and Legacy of *The Ladies' Diary* (1704–1840): A Woman's Declaration

Frank J. Swetz

MAA PRESS An Imprint of the AMERICAN MATHEMATICAL SOCIETY

Providence, Rhode Island

2020 *Mathematics Subject Classification.* Primary 01-XX, 97-XX.

Cover art by Dorothy Weir, Mechanisburg, PA. Used with permission of the artist.

For additional information and updates on this book, visit
www.ams.org/bookpages/spec-101

Library of Congress Cataloging-in-Publication Data

Names: Swetz, Frank J., 1937– author.
Title: The impact and legacy of The ladies' diary (1704-1840): A women's declaration / Frank J. Swetz.
Description: Providence, Rhode Island: MAA Press, an imprint of the American Mathematical Society, [2020] | Series: Spectrum; volume 101 | Includes bibliographical references.
Identifiers: LCCN 2020034724 | ISBN 9781470462666 (paperback) | ISBN 9781470463199 (ebook)
Subjects: LCSH: Ladies' diary. | Women in mathematics–Great Britain–History–18th century. | Women in mathematics–Great Britain–History–19th century. | Mathematics–Great Britain–Periodicals. | Women's periodicals, English–Great Britain. | Mathematics–Study and teaching–Great Britain–History–18th century. | Mathematics–Study and teaching–Great Britain–History–19th century. | Women–Great Britain–Intellectual life–18th century. | Women–Great Britain–Intellectual life–19th century. | AMS: History and biography. | Mathematics education.
Classification: LCC QA27.5 .S94 2020 | DDC 510.82/0941–dc23
LC record available at https://lccn.loc.gov/2020034724

The frontispiece shows an eighteenth-century female study group. Such gatherings discussed literature and poetry, but also, as evidenced in *The Ladies' Diary*, mathematics. Initially this illustration appeared as an introduction in the 1744 edition of *The Female Spectator*, the first British journal to be edited by a woman, Eliza Fowler Haywood. The journal lasted for two years. Source: Hathi Trust, Public Domain.

Contents

Foreword   xi

Preface   xiii

1 A Fortuitous Encounter   1

2 *The Ladies' Diary: or Woman's Almanack:* A New Kind of Publication

For those unfamiliar with this publication, just what was *The Ladies' Diary*?   7
2.1 Seeking a Perspective   7
2.2 Testimony from Users and Observers   10
2.3 What did We Learn?   12

3 *The Ladies' Diary*: Conception and Evolution

How did the editors shape and control the direction in which the *Diary* developed and progressed?   15
3.1 Almanacs   15
3.2 *The Ladies' Diary*, Shapers of its Conception: The First Three Editors   18
3.3 Sustainers of the *Diary* and Advocates of its Mathematics   30

4 "Delightful and Entertaining Particulars"—Problem Solving

Why enigmas and mathematical problems?   35
4.1 Some Thoughts   35
4.2 The Enigma, A Word Maze   36
4.3 The Enigma's Enduring Popularity   38
4.4 The Enigma in the Period of *The Ladies' Diary*   39
4.5 The Mathematical Questions   40
4.6 A Sampling of Mathematical Questions with Different Editorships   48

5 **Mathematics, Education, and Women in Eighteenth- and Nineteenth-Century England**

   What were women's opportunities to study and know mathematics?   57
   5.1 Emerging Mathematical Priorities   58
   5.2 A New Educational Thrust   63
   5.3 A Woman's Exposure to Mathematics   69
   5.4 Contemporary Testimony on the Subject of "Women and Mathematics"   75

6 *The Ladies' Diary* **as a Facet in the Mathematical and Scientific Transition of the Era**

   What intellectual value did the *Diary* possess?   79
   6.1 Fulfilling a Mathematical Need   79
   6.2 Acquiring New Mathematical Skills and Understanding   80
   6.3 The *Diary* as a Provider of Scientific Facts   85
   6.4 Instructional Essays and Dialogues   90

7 **"*Dia*" as a Mathematical Testament**

   Did the *Diary* reflect and support the mathematical reforms taking place during the span of its publication? 95
   7.1 The Winds of Change   95
   7.2 *The Ladies' Diary*: Trends and Influences   99
   7.3 Further Mathematical Trends   108

8 **Women and the *Diary***

   Did the *Diary* really serve the needs of women?   113
   8.1 Was The *Ladies' Diary* truly a Ladies' Periodical? 113
   8.2 The *Diary* as a Vehicle for Female Mathematical Expression 115
   8.3 The Ladies' Statements   117
   8.4 Prominence and Recognition   119

9 *The Ladies' Diary*, **a Noteworthy Heritage**

   So what did we learn about the *Diary*'s effects and societal impact?   121
   9.1 A Wider Influence   121
        A Family Tree of Offspring   122
   9.2 Offspring in America   124
   9.3 A Conclusion   125

Contents

Epilogue 129
    What more can we learn from *The Ladies' Diary*? 129

A Selected Word Puzzles from *The Ladies' Diary* 131

B Sample of a Complete Set of *Ladies' Diary* Mathematical Exercises 135

C Selected Problems with Worked Solutions 141

Bibliography 153

Endnotes 161

Foreword

Even though 2020 marks the silver anniversary of my entrance into the profession of the history of mathematics, I am nearly certain that I have never met Frank Swetz in person. However, we have worked together since 2014, when I joined the editorial board of *Convergence*, the Mathematical Association of America's online journal for the history of mathematics and its use in teaching that was founded in 2004 by Frank and Victor Katz. Using email to discuss the process, I have posted dozens of the several hundred "Mathematical Treasures" Frank has collected, which consist of images and explanatory text for landmarks and other works in the history of mathematics that are useful in teaching mathematics. Additionally, I have edited and posted submissions from other individuals and institutions, and I have even written a number of my own entries for the collection, spotlighting mathematical objects held by the Smithsonian Institution's National Museum of American History.

This book discusses several more of our shared interests, including the history of women and mathematics; the history of print culture, which involves interactions between authors, publishers, and readers; the history of formal and informal mathematics education; and modern British history. As Frank notes in his preface, *The Ladies' Diary* has long fascinated historians of mathematics; indeed, the sources he lists are all on my own bookshelves. One of the strengths of his addition to this literature, from my point of view, is that he has made the periodical accessible to a non-academic audience. If you have been gifted this book or picked it up wondering what a "diary" for women has to do with mathematics, you will find a fascinating cast of colorful characters in the journal's editors. You will get a sense of how rapid social and economic changes in England helped facilitate the success of *The Ladies' Diary*, and you will learn about the structure of the publishing industry in the eighteenth century. You will be introduced to several other contemporaneous and interlinked developments, such as increased literacy rates, ever-louder questions about what and how women and non-elite classes should learn, the relationship between problem-solving and a widespread audience of capable amateur mathematicians, and concerns raised by British mathematicians about what a professional discipline of mathematics should look like. The story of *The Ladies' Diary* also intersects with stories on the European and North American continents. Overall, Frank has provided a lively biography of this periodical.

The Ladies' Diary has been popular with historians of mathematics because it is wonderful to uncover a community of female readers who energetically engaged in doing and communicating about mathematics, perhaps particularly during a period in which most of the other women who are remembered for their mathematics were exceptional in their achievements and in their social and intellectual access to notable male mathematicians who could encourage and publicize their work. (If you are not familiar with names such as Maria Agnesi, Emilie du Châtelet, Mary Somerville, and Ada Lovelace, do go look them up after finishing this book.) At the same time, though, *The Ladies' Diary* poses a puzzle to historians who study women because it was under the editorial control of men throughout its history, and most of its contributing authors were men. Furthermore, during the latter two-thirds of the journal's history, male readers displaced its initial female audience. Frank points out some reasons for this, such as the appeal of introducing content oriented toward male university students into an already established publication. As someone who has repeatedly taught Joan Scott's groundbreaking article on gender and history to undergraduate history majors,[1] I also think one of the histories of *The Ladies' Diary* that is yet to be written needs to consider both its men and its women through the lens of gender. How did expectations for how men ought to behave in eighteenth- and nineteenth-century British culture influence their actions with respect to *The Ladies' Diary*? How might *The Ladies' Diary* provide a case study for power relationships among men, among women, and among men and women?

In his epilogue, Frank suggests additional directions for research by future scholars. His bibliography can serve double duty as a list of recommended sources for those readers who want to learn more about any of the many topics addressed in the story he tells. Although it has been conducted electronically, my relationship with Frank is one of my personal "mathematical treasures." I hope you enjoy his book.

–Amy Ackerberg-Hastings
–Co-Editor, MAA *Convergence*

Preface

Over the years, during my investigations into societal effects on mathematics teaching and learning, distractive topics and subjects have appeared. They were distractive in the sense that they were not directly pertinent to the immediate inquiry; nevertheless, they were often very interesting and I made note of some of them for later inquiry. So it was with *The Ladies' Diary: or The Woman's Almanack*. In the 1970s, questions regarding female performance in mathematics were in the forefront of educational research, particularly the issue of female participation in mathematics: "Why are so few women attracted to the study of mathematics? What factors in the female make-up repel them from mathematics?" Elizabeth Fennema at the University of Wisconsin led the efforts to find answers to such questions.[1] But in this process of understanding the effects of gender on mathematics teaching and learning, many researchers often came up with exotic causes: left-brain versus right-brain dominance; lack of testosterone, as mathematics is an aggressive subject; and so on. Based on my research and observations, particularly in non-Western societies, I had put the issue to rest in my mind, concluding that historical social and cultural conditioning in Euro-centric societies established intellectual limits for women, excluding them from mathematical pursuits.[2] Furthermore, I felt this aura of discrimination was deservedly lessening and would soon be extinct.

It was in this period of concern and inquiry in 1977 that I first came across mention of *The Ladies' Diary: or The Woman's Almanack*, a British eighteenth- and nineteenth-century journal that encouraged and promoted mathematics for women. This periodical was the subject of an article by Teri Perl in the *Mathematics Teacher*, a journal dedicated to mathematics teaching, published by the National Council of Teachers of Mathematics.[3] The concept of proper English ladies, in the mold of Jane Austen's or the Brontë sisters' heroines, doing mathematics and revealing it for the public review in this period of history was fascinating. I knew that British society of the eighteenth and nineteenth centuries had strict codes of conduct and expectations for their young ladies that distinctively did not include involvement in mathematics and science. Intellectual pursuits for women in general were discouraged. It was thought such strenuous cerebral activity would cause a "fever of the brain." English women of the period were not even afforded an academic education. But here was an apparent contradiction, an aberration, "How and why did it happen?" I had to find out more. In 1979, Perl published a more extensive article on the subject, "The Ladies' Diary: or The Woman's Almanack, 1704-1841."[4] Eventually, I came across the work of Shelly Costa, whose Cornell dissertation of 2000 followed up by her *Osiris* article: "The *"Ladies' Diary"*: Gender, Mathematics, and Civil Society in Early-Eighteenth-Century England", 2002, revealed the existence and significance of *The Ladies' Diary* to a broader audience.[5] Costa's scholarly study traced the evolution of the *Diary* to the middle of the eighteenth century, 1754, and noted the feminist mathematical statement it made and its place in the English "dualist civil society". In 2008, Joe Albree and Scott Brown published " 'A valuable monument of mathematical genius': *The Ladies' Diary* (1704-1840)" in the journal *Historia Mathematica* [**AB09**] in which they examined the *Diary*'s mathematical content in relationship to the mathematical climate of the period.[6] Their article strengthened my opinion even more of the intellectual and societal significance of this eighteenth-century ladies' periodical.

The deciding impetus for a further investigation on my part was supplied by John Heilbron's splendid book, *Geometry Civilized: History, Culture, and Technique* (2000 edition) [**Hei00**].[7] Throughout the text, his discussion on geometry is supplemented by illustrative problems. Many of these problems and their solutions were taken from *The Ladies' Diary*. The problems' compositions demonstrated imagination and ingenuity as well as a broad knowledge of fundamental mathematics. Simply, they were good problems! At this time, I edited a problem section for the Mathematical Association of America's e-journal *Convergence*, "Problems from Another Time", and I began to include some of *Diary*'s problems in my selections. They were well received by the viewing audience and frequently made their way into classroom teaching. When, in 2012, I published a book on problem solving: *Mathematical Expeditions: Exploring Word Problems Across the Ages*, I included a chapter of problems selected from *The Ladies' Diary*.[8] I had read and used the *Diary* as a problem resource, but then I wished to know more about it. "What is the detailed story, or stories, within this ladies' journal awaiting to be told?" "Why did the journal come into being?" "How was it received?" "What was its impact, both contemporary and prolonged?" Thus, it is with this background and questions that I begin my exploration to obtain a better understanding of *The Ladies' Diary*. Come with me!

The first chapter provides a general introduction, a feel, a foreshadowing as to where the investigation will take us. Each succeeding chapter then focuses on and explores a specific issue concerning the development of *The Ladies' Diary* or its impact. Directly under the titles of these probing chapters, a question is posed for the reader, an overriding query intended to be resolved within the immediate reading. An extensive bibliography is offered both to document the validity of the information presented and to assist future scholars who might wish to continue this research into *The Ladies' Diary*. Three appendices provide the more adventurous readers with further problem-solving challenges. I am particularly indebted to several people who assisted me in the initiation and processing of this manuscript. First, Teri Perl who, through her publications (1977, 1979), introduced me to *The Ladies' Diary*. Next, a cadre of people that assisted in preparing the manuscript for publication: Jennifer Jillson, who read the completed manuscript and made several important suggestions that enhance the final version; Dorothy Weir, a neighbor and friend, for her artistic renderings; Amy Ackerberg-Hastings [**AH08**], a colleague–historian who contributed the Foreword; and finally, Carmela Ortiz Menendez, for her coding of the manuscript into the required LaTex format. Lastly, I wish to thank my grandson Braden Howe for his encouragement. Of course, the presentation of facts and the subsequent conclusions stated rest with me, the author.

This journey of understanding we will undertake spans more than the approximately century and a half of the *Diary*'s existence, 141 issues, and, to an extent, it must probe the psychological, social, and intellectual precedents of this periodical's appearance. There is much to examine and learn. Not every issue of the journal can be read in detail, but its basic objectives and messages can be understood. This is what I will try to discern and relate. When possible, I will use the *Diary*'s actual words and illustrations to convey meaning as it was intended. In doing so, I will expose the reader to some of the customs and foibles of pre-Victorian British society: word usage, spellings, and punctuation. When possible, I will quote witnesses' testimony of the time: correspondents, editors, and critics. This practice is intentional, as I believe the best way to understand an historical situation is to culturally participate in it as fully as possible: attempt to recognize the wit, the verbal ostentation and posturing, the format, punctuation and emphasis of written statements, and the intellectual motivation. Feel the stirrings of social change that are beginning to surface and project them into the formation of the British society you know today. Let us start the journey.

–Frank Swetz

1

A Fortuitous Encounter

Two nattily dressed gentlemen briskly proceeded down Chertrey Road before turning into St. Michael's Alley. Nigel Wallingford and Geoffrey Todd were school chums at Ashley Academy for young men. The year was 1754. Nigel was now apprenticing at his father's trade house. The elder Wallingford had inherited the family import business which in recent years was conducting a lucrative trade with the East Indies. Hopefully, Nigel would take over the business. Young Mr. Todd was reading law with a respected London solicitor. Moving through the London fog and the smoke of coal-fed evening cooking fires, the men resembled vague specters: the lawyer-in-training held a white handkerchief to his nose and mouth, protecting himself from the acrid stench of the nearby River Thames; Nigel sported a rosewood walking stick, a gift from his father and an object for protection in case of an unfriendly encounter. Crafted in India, the walking stick was one of the many new and excellent products produced in Great Britain's far-reaching Empire. Haloed light from a mist-shrouded lantern marked the entrance to their destination, the Black Turk Coffeehouse. They entered into a smoke- filled room almost as opaque as the outside alleyway, but they found the warm aroma of tobacco, the smell of brewed coffee, and the din of boisterous conversation welcoming. As they hallooed a greeting into the establishment, a faceless voice shouted out "What news, Sir?" a query to which Nigel responded, "The Schooner, Mary Johnson, with Master Henry Orlip, made harbor today with a full cargo of cotton and tea from the far Indies," an announcement that readily promoted the Wallingford business but was also of interest to the many merchants-clients at the coffeehouse. London coffeehouses were known as centers for the dissemination of business information and the undertaking of financial transactions.[1] See Figure 1.1. The two men had come to the coffeehouse to meet a fellow

1

classmate, Elsmont Potter, who had gone on to university at Cambridge and had come down to London for holiday. "Hi Hoo Ashley, Wally, Geoff, over here," the call from their friend "Els" guided them through the grey smoky mist to a distant corner.

He sat alone in a booth. A single candle illuminated the paper-strewn table before him. Els greeted his friends warmly and ordered two bowls of the "Mohammedan brew," Turkish black coffee, to be brought to them.[2] Nigel balked at the "syrup of soot," preferring the more fashionable drink, tea; however, in the

Figure 1.1. London coffeehouse of the period, witnessing an "in-your-face" argument. From: Ned Ward, The Coffee House Mob, frontispiece to Part IV of *Vulgus Britannicus, or the British Hudibras* (London, 1710). Source: Wikimedia Commons.

spirit of friendship, he accepted the offering and sipped at it cautiously. "I say, what are you reading so intently?" Geoff exclaimed as he examined the open magazine in front of Elsmont. "Gad Potter, *The Ladies' Diary*, is this what university has done for you, changed your reading habits to those of the fairer sex?" While half in jest, Geoff's comments did indicate a perplexed concern. "Rest easy, my friend, these are good maths problems here. You should try some," Els challenged. To which Goeff deflected: "No, you sound like Wormely, our old maths master at Ashley. You know how I always hated maths, the numbers, and doing problems. It's the stuff of tradesmen." Reading the problem Ham set before him, he gaffed further, exclaiming: "Couriers, indeed... This problem is proposed by a woman!"(She signed herself Miss Maria A-t-s-n.) To which Nigel remarked "By Jove, the impertinence; she is probably a Philomath, one of those bluestocking women. Soon they will be even coming into the coffeehouses. These women are shameful." Geoff supported his colleague's antifeminist protest, adding, "She is probably a prune, a dried up old spinster." Elsmont laughed at the bluster, assuring them: "Whoever she is, it's a damn good problem and I'm going to send her a solution."

The problem's proposer was a Mary Atkinson and her problem was:

There are three cities A, B, and C, lying in the same road; whereof the first is 136 miles distant from the second, and the second 104 distant from the third: From A to B a courier travelling two days journey; and from B to C two days more, diminishing his distance every day alike, from first to last. What miles did he travel each particular day?[3]

Els sent his answer to this problem to the *Diary*, as required, but also, in curiosity, addressed Mary directly, asking her age. His submitted mathematical solution was correct and "the woman" responded to his inquiry in the next issue of the *Diary* by answering the question by means of an arithmetic riddle:

Five times seven and seven times three

Add to my age, the sum will be

As many above six nines and four

As twice my years exceed a score[4]

Elsmont unraveled the word puzzle to determine that Miss Atkinson was eighteen years old. Upon this discovery, he patronizingly counseled her: "A very good age for matrimony, Miss."

This scenario, this encounter, part fiction, part reality, is very telling. The proponents: Nigel, Geoffrey, and Elsmont are all fictitious players in a real situation, fleshed out to fit their roles. The tenor of the encounter reflects the social and intellectual climate of the period. Mary Atkinson did, however, exist and proposed the stated problem to *The Ladies' Diary* in the year 1754 and gentlemen of this time were not supposed to have a particular interest in mathematics. Among male readers who solved the problem there actually was one, an "E.P." who asked her age and when finding it, via her riddle response, indicated his disapproval of such female daring and advised her to get married. What does this episode tell us?

In the late eighteenth century, Britain was beginning to benefit from the dividends of the Industrial Revolution and the spoils of its Imperial Empire. Commerce was booming, trade was flowing daily into the ports of London and Liverpool. Increased wealth had spawned an expanding middle class with a social division dominated by the "Gentlemen," a seemingly privileged, male-orientated fraternity, made up of well-dressed men who gathered in clubs and coffeehouses. Perhaps reflecting the examples of their aristocratic peers, this new social stratum insulated itself within British society by adhering to strict social codes and expectations. A gentleman or "gent," even though his family background may have included skilled craftsmanship or mercantile pursuits, shunned physical work or approximate commercial involvement. Intellectual endeavors were also viewed with some disdain thus Geoff's rejection of mathematics. Even when more liberal university entrance standards were put in place, they did not attract young men to the institutions. The female counterpart of the gentleman was a member of the "Fair Sex," the "Lady," the daughter from a respectable family. This category included the offspring of clergy, academics, professional people such as lawyers, country squires, if their land holdings were large enough, and even merchants of admirable social standing were worthy of note, provided of course that their wealth was sufficient enough to support an attractive dowry. The expectations for ladies were shaped by the ideals and fantasies of the gentlemen. Women were to be wives and the bearers of children, to run a household with the assistance of at least one servant, to be able to entertain guests in a pleasant manner and, in general, to be a passive ornament for their husbands. They were expected to have some education, or rather, training, in the arts and graces of being a lady which primarily included the possession of a limited intellectual curiosity and perspective. Ladies' opinions, if any, were to be those of their husbands. Women were to be modest. Certainly, they should not be involved with the sciences or mathematics. Benjamin Martin, noted eighteenth-century scientific lecturer and publisher, commented on this situation in 1755 [**Mar55**]. In his illustrative dialogue between a brother and sister, the brother notes how it has become the fashion for women to study philosophy [the sciences], to which the sister tellingly responds:

> I often wish it did not look so masculine for a Woman to talk of Philosophy in Company; I have often sat silent, and wanted resolution to ask a Question for fear of being thought assuming or impertinent. I should be glad to see your Assertion verified; how happy will be the Age when ladies may modestly pretend to Knowledge, and appear learned without Singularity or Affectation![5]

At the center of this genteel society composed of ladies and gentlemen were the conditions and amenities of leisure. These individuals were the product and beneficiaries of the recently expanding state of leisure—the freedom from physical or demanding work. London gentlemen attended their clubs, frequented the coffee or tea houses where some business might be conducted, gambled, enjoyed the races, and were familiar with the brothels of the great city. Country gentlemen ran their estates, hunted, and visited neighbors. As expected, ladies attended their husband's house, hosted teas, bore and mothered children, and demonstrated such acquired feminine skills as watercolor painting, embroidery, or playing the piano and singing—all to the approval of their husbands. Literacy rates had increased in England both due to and encouraging the printing industry. A variety of broadsheets, journals and almanacs, expanded sources of knowledge, were published for the enjoyment and information of ladies and gentlemen.

But not all the socially uplifted population heeded the societal and intellectual constraints imposed upon them. Opportunities for self-education, including the increased availability of printed information, opened new fields of learning and understanding for many. Subjects previously limited to philosophers, clergy, and academics now became available to the broader audience. New ideals of individual worth and social organization were introduced. Those who flocked to this available learning and knowledge were often called *"philomaths,"* lovers of knowledge. Knowledge served as a stimulus for reforms. In some sectors of society, individuals and groups became dissatisfied with their existing prescribed social, intellectual, or economic status. They sought change. In particular, some women began to reach out for more independence and freedoms, among which were the ability to arrive at their own intellectual opinions and modes of expression, and to be heard and be part of the larger societal discourse. Such women were often labeled "Bluestockings," a term that was frequently employed in a derogatory sense, as it was felt the Bluestockings impinged on the intellectual prerogatives of men.

The learning and doing of mathematics was deemed inappropriate for young ladies and women, in general. Mathematics, if pursued, was a male activity and even then, limited. In defense of such an assertion against women, the complexities of calculations were often said to be harmful, strenuous and taxing, to the delicate female brain. Nigel and Geoff's critical comments to the fact that a lady had submitted a mathematics problem to a journal and further was brazen enough to attach her name, albeit disguised, to the offering was based on such a belief. But, yes, women at this time were doing mathematics in England and, a popular almanac/journal, the *Ladies' Diary* established in 1704, provided a forum and stimulus for their mathematics. This periodical, the first of its kind in England [and the rest of the Western world], became extremely popular and exerted a broad and profound influence that lasted well over a century.

Both of these phenomena: women actively and openly participating in mathematical activities at this period of history despite the cultural and social constraints mitigating against them, and, the existence of an actual journal, *The Ladies' Diary: or, a Woman's Almanack*,[6] that encouraged and supported their mathematics, are worthy of a closer examination, analysis, and further discussion.

2

The Ladies' Diary: or Woman's Almanack: A New Kind of Publication

For those unfamiliar with this publication, just what was *The Ladies' Diary*?

2.1 Seeking a Perspective

From the mid-seventeenth century onwards, as Great Britain began to benefit from the dividends of the Industrial Revolution, literacy rates increased along with the existence of leisure time. The reading British public eagerly consumed news and information provided by a variety of print sources: almanacs, magazines, journals, broadsheets, and formal newspapers.[1] For the British reading audience almanacs were the most desirable, affordable, and convenient forms of reading material. The estimated official sales record of these books for the year 1801 was 605,000; by 1839 the total had reached a record of 694,000.[2] Many more were produced and sold illegally, that is, plagiarized, copied imprints of an existing work, usually without the required government tax duties paid. Considering that the population of Great Britain in 1850 was about 25 million people, of which *approximately* half were literate, and that a periodical such as an almanac was most likely shared by several people, for example, five individuals, would indicate that approximately a majority of the reading population were in direct contact with almanacs.

Writers and publishers readily satisfied the literary demands of the rising consumer society. One such entrepreneur who saw an opportunity to improve his financial status and satisfy a perceived popular need was John Tipper (1663–1713), a Coventry schoolmaster. He would publish an almanac that was different: it would be designed for women, avoid political and religious issues, and promote a contemporary, scientific world outlook. Tipper would purposely disdain astrological references, features then prominent in most existing almanacs. He would not make predictions and instead provide information on recent discoveries. In seeking a unique niche for his publication, the schoolmaster was mindful of the growing leisure class of women and their restless intellectual state. But his was not the first popular periodical in England to be published for women. In 1691, two monthly magazines appeared: *The Ladies' Mercury* and *The Ladies' Journal*, both conceived and brought forth by almanac editor and publisher John Dutton. These magazines moralized and preached to women. They lacked appeal and both died within a short period of time. In order to publish his almanac, John Tipper had to secure the permission and support of the London printers' guild, The Worshipful Company of Stationers (a group of ten individual printers together with their bookbinders, engravers, font and paper makers, and book sellers) held a royal monopoly on the printing and distribution of almanacs. The Company of Stationers was aware of Dutton's past failures but they approved Tipper's proposal for a periodical devoted to women, cautiously limiting a first issue to three thousand trial copies and fixing the price at three pence, more expensive than competing almanacs that sold, at that time, for two pence an issue.

Tipper's conception, *The Ladies' Diary: or, a Woman's Almanack* was first published in 1704. The new periodical was only about three and a half inches wide and six inches long and in book form, containing some forty pages. Other, similar publications were equally small—possibly reflecting an economy in the use of paper but also offering convenience of transporting and carrying whether in one's pocket or purse. "Pocket books" were becoming fashionable. The cover contained a portrait of the reigning queen. Besides bearing the extended title: "*The Ladie's Diary: or, the Woman's Almanack*", then identifying the respective, "Year of Our Lord", for which it had been compiled, the title page offered an enticing promise: "Containing many Delightful and Entertaining Particulars, peculiarly adapted for the Use and Diversion of The Fair-Sex". See Figure 2.1. An issue of the *Diary* was divided into several parts: initially under Tipper's editorship, a lengthy "Preface" paid varied tributes to women in general; a calendar, with separate pages for each month, and meteorological information—astrological predictions were avoided. It also featured receipts [recipes] and medical advice; discussions on love and marriage; entertaining stories and poems and enigmas, or word riddles. Readers were asked to send in their solutions to the enigmas and correct puzzle solvers would be acknowledged in the next annual issue of the periodical. To Tipper's satisfaction, and perhaps to the amazement of the Company of Stationers, *The Ladies' Diary* proved to be a great success. Almanacs for the following year were usually released in November, on Almanac Day, so that readers were prepared and could plan for the coming year. In the second year of the *Diary*'s existence, 1705, the Company of Stationers printed four thousand copies that were completely sold out by January.

A correspondent to the 1707 *Diary* sent in two arithmetical puzzles or problems. Tipper printed them and repeated the problems in 1708. Letters inundated him from his female readers requesting more mathematical exercises. He complied in the following issues, establishing a format that would survive the life of the journal. *The Ladies' Diary* would evolve into a problem-solving periodical consisting of three principal parts: a calendar and meteorological information, mathematical problems, and enigmas.

Figure 2.1. The cover of 1709 the issue of the *Diary* with a picture of Queen Anne. Source: Harry Ransom Center, University of Texas, Austin.

John Tipper and his two, immediate, succeeding editors, Henry Beighton (1687–1743) and Robert Heath (1720–1779), established a trajectory for *The Ladies' Diary* that would carry it into the next century, a journey supported by a series of diligent editors. By 1718 the *Diary*'s output rose to nearly seven thousand and by the middle of the century sales records reached about thirty thousand issues per annum.[3] Such a demand was very impressive. In the highly competitive field of British almanac publishing, *The Ladies' Diary* became and remained, until its demise, a renowned success.

2.2 Testimony from Users and Observers

The Ladies' Diary sold well but what was its real worth? How was it used and reflected upon by its readers and subscribers? What did they think about it? Fortunately, there does exist some contemporary testimony in the form of "Letters to the Editor" and reviews that speak to the reception and utilization of *The Ladies' Diary*. This information is spoken and written in the manner that places it in its time, eighteenth- and nineteenth-century England.

The 1718 issue of the *Diary* contains the following "Letter Commendatory" written by a Mr. Meredith Jones in November of 1716. Apparently, Mr. Jones was a schoolmaster:

> Sir,
>
> When I reflect upon the extreme Usefulnes of your incomparable Diary, I can't be but mightily pleas'd with so good and public Advantage you are pleased to communicate by your ingenious Industry to our Nation. I don't affirm this by way of Complement, but as it is real Truth, as I can testify, not only in my self, who by it have been egg'd and induced to have a Love and Esteem for that incomparable Analytic Art; but also as I have experimentally found in others, as I was, by Providence, plac'd in a public Station in the World to teach Youths in that and the like Sciences. I know this Day many eminent Men that owe (as I have heard them to confes it) their Knowledge in those Mysteries to the ingenious Mr. Tipper, deceas'd; I mean owe it so far, as his *Diary* was the first Occasion of applying themselves that Way: And I don't doubt but it will have the like good Effects upon other Genius's in future Ages, that might have lost the Advantages of doing Good to their Country, by unproving such an universal advantageous Knowledge. ...[4]

John Tipper had died in 1713. Jones goes on in his correspondence to offer suggestions for the submission of mathematical problems. While Meredith Jones's comments are mainly focused on the "incomparable Analytic Art," mathematics, his compliment as to "Usefulness" has a wider scope. *The Ladies' Diary* was promoted and sold as an almanac and, as such, provides the appropriate referencing services and knowledge expected from by an almanac: dates, seasons, times and special days of note. The calendar pages were heavily annotated with supplementary information, increasing the reader's worldly awareness. See Figure 2.2.

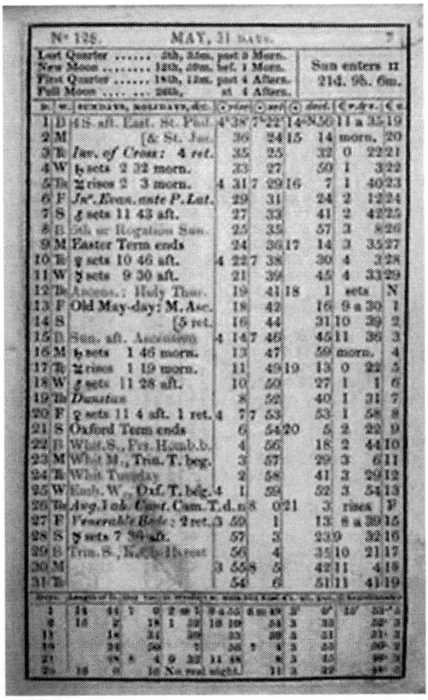

Figure 2.2. An informative calendar page from the 1831 *Diary*. Rights: Public Domain, Google-digitized.

A correspondent in 1726, Richard Whitehead, laments the fact that some people dismiss the *Diary* as it is small in size and is merely an "almanac". He urges that:

> ...such persons ought to be told, *First*, of the extensive Uses of Mathematical Learning, and of the infinite Advantages they are to Mankind, that the Study is not only delightful, but quicken the Invention, strengthens the Judgment, and enlarges the intellectual Faculties. *Secondly*, That in the *Diary* there has been exhibited a great Number of difficult, curious and useful Questions, in all Branches of the Mathematiks, not to be found in any Author, with the best Methods of solving them, which in any other Method probably would never appear to the World. *Thirdly*, That the *Diary* has incited and led many Persons to the Study of Mathematicks, who otherwise perhaps wou'd not have turn'd Their Thoughts that Way, and has also exercised those who have before studied them so as not to forget what they learnt, and by exciting an Emulation whereby they extend their Mathematical Knowledge. ...

Whitehead goes on to comment favorably on the abilities of the problem solvers and ends his letter by noting:

> The solving of some Questions does take up much time, and exercises as well the Hours pleasantly, innocently and profitably, with inconsiderable Charge is highly commendable: especially since Time is generally foolishly and expensively thrown away.[5]

He, himself, was a consistent and proficient problem solver.

Furthermore, a noted librarian and antiquary of the times, Sir Henry Ellis, described *The Ladies' Diary* as:

> ...a work, while humble in its beginning, has exerted a great and beneficial influence upon the state of mathematical science in the Country for nearly a century and a half.[6]

Thomas Kirkman (1806–1895) was one of England's notable, nineteenth-century, "research mathematicians." Particularly active in the field of combinatorics, he originated the "School Girl Problem"[7], published in the *Diary*'s successor in 1850 but writing in 1849, he commented:

> I confess it to be my belief, from a limited observation of graduates and non-graduates, that when the difference between prizes awarded by the authorities on either side is considered, an incomparable greater share of glory of kindling and cherishing a pure and lasting love of mathematical science in *men* as well as *boys*, must be attributed to the immortal lady *Dia*, than all the universities and colleges of these kingdoms put together, to all our Lyceums, Athenæums and Philosophical Societies, and to all our Imperial Boards of War and Peace.[8]

Perhaps Kirkman is overly enthusiastic and a bit unrealistic as to the mathematical impact that *Dia* (an affectionate nickname for *The Ladies' Diary*) created but this enthusiasm is telling, as is his limited reference to men and boys, considering it is a ladies' journal. Nevertheless, he, as an experienced and highly respected mathematician, certainly expressed a high opinion of *The Ladies' Diary*. Previously in 1808, another eminent mathematician, John Playfair, had issued similar praise on the status and influence of *The Ladies' Diary*.[9] So, in its time, those who came in contact with the journal considered it a unique publication. *Its mathematical problem solving challenges seem to appeal to a largely male audience.*

2.3 What did We Learn?

From these selected comments of people directly involved with *The Ladies' Diary*, a few conclusions can be deduced and will serve as issues for further investigation, among which are: the periodical was popular with the public and it was appreciated for the information it conveyed; *The Ladies' Diary* served an educational purpose; through its feature of "Mathematical Questions" requiring problem solving, it promoted the use and learning of mathematics. This mathematical section was initiated by a request from female readers who at the outset strongly participated in this area of problem solving. At this time in history, there seemed to be a mathematical ferment/discontent going on in England, a movement in which *The Ladies' Diary* was playing a central role. All participation in mathematics, as described so far, seems limited to men. No mention of women doing mathematics is made even though much of the discussion concerns *The Ladies' Diary*, a journal for women! Thus three "forces" or stimuli emerge that influenced and controlled the course of *The Ladies' Diary* over its lifetime:

- The succession of editors.

- Ladies, for whom the journal seemed intended.

- British mathematical reforms and movements.

Each of these categories will be examined in some detail.

3

The Ladies' Diary: Conception and Evolution

How did the editors shape and control the direction in which the *Diary* developed and progressed?

3.1 Almanacs

To a modern reader, captive subject to a variety of electronic news media, the concept of an almanac may seem quaint, even primitive, but it should be remembered that since the invention of writing, humans have always been recording and consulting celestial information that they believed had an effect upon their lives. Such astronomical and astrological data formed the basis of written records that eventually would be called almanacs. Almanacs became calendars and chronological handbooks that supplied information such as the phases of the moon, eclipses, appropriate times for planting crops or beginning a voyage; and provided charts, horoscopes, giving auspicious and inauspicious times for various daily activities. They allowed for prognostication. In a limited sense, they predicted the future. Almanacs, usually issued once a year, provided useful guides for the planning of daily activities from arranging a marriage to buying a horse. With the advent of moveable-type printing in Europe, one of the first printed publications by the Gutenberg Press was an almanac in 1457. In England, from the sixteenth century onwards, almanacs found a ready market, selling very well.[1] In a royal charter of 1557, King James I gave the London printer's guild, The Worshipful Company of Stationers, a monopoly on the printing and distribution of almanacs on condition that they provide an annual annuity to support Oxford and Cambridge universities. The Stationers Company, the initial printers of Milton and Shakespeare, found the arrangement most agreeable, providing them a source of steady income.[2] Usually priced at a penny or two, these broadsheet almanacs, later to become little books, were within the purchasing power of the common folk, provided they were literate. Agriculture and maritime trade was becoming the mainstay of the English economy. As a result, any source of information on the flow of tides and weather predictions was important to trade and businessmen just as advice on planting, weather, and harvesting times was relevant to yeoman farmers; almanacs supplied this data. For hundreds of years, mariners and merchants had kept their own notebooks/diaries of technical and general information pertinent to their specific professions. Now, in a sense almanacs provided a great portion of this information. Many people with superstitious tendencies, probably the majority of almanac readers, purchased almanacs simply to obtain their personal astrological predictions—guidance for daily affairs.

At first the writers of European almanacs were practitioners of the relative disciplines involved: astronomy/astrology or mathematics. Such historically renowned scientists as: *Regiomontanus* (1436–1476) credited as compiler of the first modern almanac on the continent in 1472, Adam Riese (ca. 1489–1559), Peter Apian (1495–1552) Girolamo Cardano (1501–1576), Leonard Diggs (1510–1558), Tycho Brahe (1546–1601) and Johannes Kepler (1571–1630), all published almanacs to supplement their incomes. The notorious *Nostradamus*, Michel de Notre Dame (1503–1566), broadcast his ominous predictions via almanacs. This lucrative publishing situation also attracted ready entrepreneurs, lacking in any particular scientific background, but lured by the profit and skilled in compiling information and employing efficient marketing practices. While building upon the basic appeal of calendric reckonings and tabulation and to attract even larger audiences, almanac writers now expanded the contents of their offerings to amuse and entertain and included puzzles, political commentary, medical remedies, and advice such as "How to court a marriage partner." Almanacs provided popular reading material for people of limited resources who could not afford proper books.

Also in this time, when the state of medicine was primitive, self-knowledge, shared information and home remedies concerning health and sickness, were eagerly sought. John Woodhouse, a major almanac publisher in the seventeenth century, offered the following advice on personal hygiene and health in his 1642 annual:

> Comb your head the hair backwards, it purgeth rheum and
> cleareth the eyes. Wash behind your ears with cold water, an
> enemy to toothache, wash hands often, feet more seldom, head

not at all. A bath should be taken at least an hour after rising,
after taking exercise, and only when the Moon was in Libra or
Pisces.[3]

Note the astrological counseling even associated with the act of taking a bath. The belief that each part of the human body was affected by one of the twelve signs of the zodiac was popularly held during the early part of the seventeenth century. Most almanacs supported this belief by supplying a reference diagram of a "Zodiac Man." See Figure 3.1.

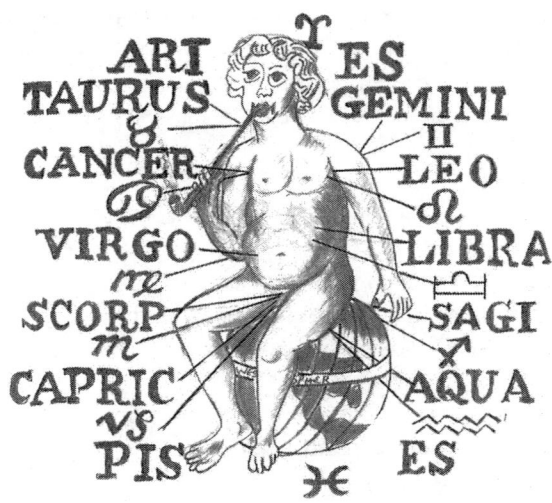

Figure 3.1. The zodiac man, a familiar diagram found in almanacs of the period. The practice of associating parts of the body with signs of the zodiac began in the Middle Ages and was evidenced in many cultures. Most British almanac illustrations for the zodiac man were more anatomical, showing the picture of a man cut open revealing internal organs which are referenced. The practice of including the zodiac man in British almanacs was discontinued in 1828. "Where does it hurt?" Pencil sketch by Dorothy Weir, Mechanicsburg, PA. Used with permission of the artist.

Eventually, the anatomical display of a naked body was considered "indecent." Such anatomical diagrams, even if not erogenous in nature, were removed from later almanacs. As for some specific medical advice, consider the receipt for a prescribed lotion intended to remove unwanted skin spots as given in the 1705 edition of the *Diary*; the mixture contains among other ingredients: "Benjamine, (benzoin, a balsamic resin); Storax (fragrant balsam); rectified spirit of wine and warm horse dung."[4] A compliant reader preparing a facial may wonder how warm and recent the horse's offering should be.

It was such astrological advice and medical guidance that John Tipper detested and would try to avoid in his planned almanac but the face remedy must have been deemed desirable for subscribers.

3.2 *The Ladies' Diary,* Shapers of its Conception: The First Three Editors

In 1699, John Tipper assumed the position of headmaster at the Bablake Hospital School, Coventry, at a salary of 20 pounds per year. This educational institution for young boys was one of the oldest in England, able to trace its origins back to a land grant by Queen Isabella, wife of King Edward II, in 1344. But despite the prestige of its age, by the time Tipper became headmaster, the school had fallen on hard times and enrollments were meager. The new headmaster modernized the school's curriculum by moving away from an emphasis on Classical Latin and Greek to include astronomy, trigonometry, surveying, and navigation. Still, he found his annual salary wanting and sought further various means to support his family. Among these options, Tipper provided private tutoring in mathematics, accounting, dialing (the geometric construction of sundials), surveying, and music. Aware of the available market for popular informational reading material, and seeking additional sources to support his family, and even to advertise his school and his personal services, the teacher compiled an almanac. He conceived of a publication devoted to women, *The Ladies' Diary: or Woman's Almanack*. Tipper believed that the *Diary* would cater to the "Fair Sex" by supplying "genteel" subject matter such as household tips, recipes, health advice, poems, and romantic stories and recognize and promote women's interests along with the standard calendar and chronological reckonings. Anna Miegon in her 2008 study of *The Ladies' Diary* premised that John Tipper modeled his almanac on the *Gentleman's Journal* (1692–1694) that despite its title, and short run, was very popular with ladies.[5] Indeed, the format and organization of the two periodicals are comparable. In his correspondence with a childhood friend and recognized London scholar, Humfrey Wanley (1672–1726), Tipper describes his sincere desire to devise a periodical "useful and appealing to women".[6] Wanley would remain his intermediary, passing on information, purchasing books and serving as a contact with London resources such as Edmund Halley, the famous astronomer. While "intended to improve the mind" by being informative, this new publication was also organized to entertain and provide diversion. Tipper included enigmas—puzzles—to be solved. In his first edition, he gave a definition of an enigma "as an ingenious and beautiful obscurity of the plainest things when discovered strikes the soul with admiration", thus an intellectual challenge and a source of satisfaction and pleasure.[7] This feature, riddles, would ultimately provide a fateful and historic direction for the almanac.

Readers were required to submit their solutions and direct queries to the editor. The journal encouraged interaction between its readers and the editor. "Delightful stories" believed to be of interest to women, some titillating by the standards of the eighteenth century, were included and sometimes serialized. For example, the 1709 edition contained such enticing literary diversions as:

"The Unfortunate Lover: His comical Description of the Beauty of his Mistress."

"A Contest between the Country Dames, City Wives and Court Ladies: Which is the Happiest Life?"

"Whether it is better to Marry for Beauty or Money?"

"Of a Wife, The Blessings of a Happy Couple."

Tipper also planned for complimentary copies of the new publication to be distributed to selected "Gentlewomen" and, if possible, Ladies of the Court. A formal proposal for the almanac was submitted to the Company of Stationers, as required, approved, and the almanac was published in 1704. For this tentative first issue, John Tipper would receive no compensation other than a promise that if the almanac went into a second year's printing, he would be given one hundred free copies to sell as he wished, keeping the revenues for himself.

As editor and major author/contributor, John Tipper attempted to fulfill his claim of appropriateness. It should be noted that this was a man deciding the "appropriate" reading material for a woman, a lady, customary within the norms of the society and his own perception of just what interested women, a phenomenon then accepted as natural and ordinary. In his correspondence with his friend Wanley, Tipper confided that he had even avoided informing or consulting his own wife on the conception and substance of the almanac.[8] Through his promotions, the schoolmaster left no doubt that his publication was intended for the Fair Sex. He constantly cultivated a female audience with flattery and pompous prose. See the excerpt given in Figure 3.2. Ladies found the publication both entertaining and useful, while men first approached it with curiosity, as a novelty, but eventually took its contents, especially its problem situations, more seriously.

John Tipper's personal expertise as an author for the *Diary* seemed to focus on supplying informative discussions on astronomical and meteorological significance and the composition and posting of the enigmas. The riddles were presented in verse and exhibited an imaginative and creative wit. Verse was "an accomplishment every young person was expected to try their hand at."[9] A reward was offered for the correct answer as well as recognition for such ability by having one's name listed in the following edition. For example, an enigma from the 1709 almanac is as follows:

The World my Age doth scarce *exceed,*

I'm old therefore you'll say indeed;

A Wonder strange it seems to be,

All Mortals should have love for me:

For I one Brother only have,

Who's hated by them as the Grave:

That's able for to tell my Name.[10]

The answer is "Abel." Many such enigmas had biblical or mythological associations, adding a further intellectual dimension to the puzzle but also quite in accord with the classical and moralistic tenor of the time. Some later enigmas were rendered in Latin, limiting the audience of participation but also hinting at the social class expected to respond—educated middle- and upper-class readers. Since Latin was a required school subject for young men, placing them at a distinct advantage over young ladies, correspondents soon complained and this practice became limited.

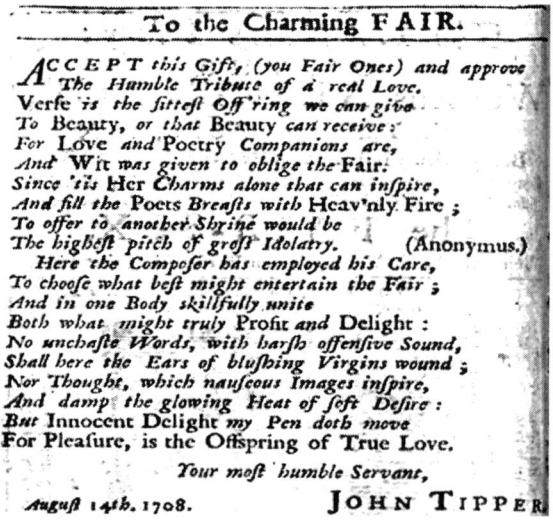

To the Charming FAIR.

ACCEPT this Gift, (you Fair Ones) and approve
The Humble Tribute of a real Love.
Verse is the fittest Off'ring we can give
To Beauty, or that Beauty can receive:
For Love and Poetry Companions are,
And Wit was given to oblige the Fair:
Since 'tis Her Charms alone that can inspire,
And fill the Poets Breasts with Heav'nly Fire ;
To offer to another Shrine would be
The highest pitch of gross Idolatry.   (Anonymus.)
    Here the Composer has employed his Care,
To choose what best might entertain the Fair ;
And in one Body skillfully unite
Both what might truly Profit and Delight :
No unchaste Words, with harsh offensive Sound,
Shall here the Ears of blushing Virgins wound ;
Nor Thought, which nauseous Images inspire,
And damp the glowing Heat of soft Desire :
But Innocent Delight my Pen doth move
For Pleasure, is the Offspring of True Love.

        Your most humble Servant,

August 14th. 1708.   JOHN TIPPER

Figure 3.2. An example of Tipper's endearing prose for the ladies. Introduction page, *The Ladies' Diary*, 1709. Source: Harry Ransom Center, University of Texas, Austin.

Tipper reported that two of his enigmas from the 1706 almanac were answered by several ladies and a certain "Mr. John White of Butterly in Devon" who then sent a pair of his own proposed "arithmetical enigmas":

(1)   In how long a time would a Million of Millions of money be in telling supposing one hundred Pounds to be counted every Minute, (without any intermission day or Night, Sunday or Work-Day) till be told?

(2)   If to my Age there added be
          One Half, One Third, and three times Three;
          Six score and ten the Sum you'll see,
          Pray find out what my Age may be?[11]

In his edition of 1708, the school teacher/editor repeated the first problem and to it added four other mathematical problems, all given in verse. Furthermore, he requested that all correspondent-supplied mathematical problems be composed in verse, as it was gentler to the ear. The *Diary* now contained a new section "Arithmetical Questions" for which answers or solutions were sought. This new feature was met with much enthusiasm as he related in the next issue of the journal in 1709. Tipper's accommodating reaction in revising the contents of his almanac would prove a fateful decision:

> Having observed by a multitude of Letters I have receive *Arithmetical Questions*, above all other Particulars, give the greatest Satisfaction and Delight to the obliging Fair, I shall in the *Diary* insist the Longer upon them, and for that Reason defer the *Receipts of Cookery*, &c. to a more favourable Opportunity: And First of the Nature.

> I have received several *Arithmetical Questions* which are very unfit for this place; my design being not to puzzle, but to *Please*; not to *perplex the Understanding*, but to *exercise the* wit, and a moderate knowledge in *Numbers*; and therefore those who are pleased to send me any *Arithmetical Questions*, I desire they may be very *Pleasant*, and not *too hard*; and likewise That they may be proposed in *verse*; which will still be the more taking among the Ladies.[12]

Simply, his readers of the fair sex preferred some arithmetical/mathematical problems, and henceforth, Tipper would supply them. Female readers of *The Ladies' Diary* now changed its intended thrust to a more cerebral and challenging involvement, mathematics. Furthermore, beginning with the 1710 issue of the *Almanac*, one mathematical problem judged the most difficult or complex would be designated as "The Prize Question." The correct solution would be published and its author would receive a material reward: complimentary copies of the periodical. An additional competitive feature was now available for the readership.

Especially noteworthy was the posting of complete solution processes for each problem in the following edition of the journal providing a learning-feedback. Readers could appreciate and evaluate different solution approaches. The first "Prize Question" of these new mathematical challenges was dated May 1, 1709, and was composed and sent to the *Diary* by an anonymous male contributor:

Dear Friend

I make bold to send you a line

T'inform you what hapt to me this day:

As I pass'd with some friends thro'Cheapside, in our way

We were viewing Bow-Steeple, says a spark that stood by,

Can you tell me sir by art; how many feet that is high?

"I'll lay you I can Sir, a piece to be spent."

We laid down our money. The sun shining plain

I measured the shadow, which I found to contain

Two hundred fifty-three feet, half a quarter,

And the clock just struck twelve as I finished the matter.

Now (Good Sir) inform me, How high is the steeple?

For you can't beat it into my head with a beetle

How is it to be done: - Were the wage to find sir,

A pretty plump girl, or a good glass of wine, sir,

I think I could do it as well as the best;

But these crabbed hard numbers I ne'er could digest.

Fail me not in this punch, Sir, whatever you do,

If you should, my dear money away I should throw:

Besides, all my friends, Sir, will laugh at me too.[13]

Although this question is directed at a male ("Dear Sir"), a lady, who signed herself as "Mrs. Mary Wright", solved this verbose trigonometric query. Mrs. Wright won the prize. Such a format of "Prize Problems" and a system of rewards would continue throughout the life of the *Almanac*; by 1711, the feature of "Prize Questions" would be extended to include the solving of paradoxes concerning geography.[14]

The editor's reception of correspondents proposing problems and enigmas increased so encouragingly that in the 1711 *Ladies' Diary*, Tipper announced his intention of publishing a new monthly periodical made up of this overflow. Entitled *Delights for the Ingenious*, a title not original on the English publishing scene, this additional literary endeavor did not last a year.[15] John Tipper died in 1713.

Table 3.1. Editorship of *The Ladies' Diary*

Editor	Tenure
John Tipper	1704–1713
Henry Beighton	1714–1743
Elizabeth Beighton	1744
Robert Heath with assistance of	
Elizabeth Beighton	1745–1753
Thomas Simpson	1754–1760
Edward Rollinson	1761–1773
Charles Hutton	1774–1818
Olinthus Gregory	1819–1840

A series of other *Diary* editors would follow and each, directly or indirectly, would affect the tenor of the journal for ladies. See Table 3.1.

The first of these editors to be chosen by the Company of Stationers was Henry Beighton (1687–1743), a surveyor and engineer from the nearby village of Griff.[16] Beighton knew John Tipper personally. Certainly, he was familiar with *The Ladies' Diary*, being the prize winner of the second mathematical challenge problem (1711). Beighton was a proclaimed "Newtonian" and promoter of the Enlightenment, a believer that the universe is governed by mathematical laws. Remember that Isaac Newton (1643–1727) was still alive at this time and the British nation was reveling in the glory of his scientific discoveries and reputation. Beighton was suspicious of the majority of magazines and journals in their ability to truthfully report factual information and their association with astrology but he respected the reputation and influence of *The Ladies' Diary* as a valid source of scientific information. He accepted its editorship. As its editor, he would, however, remain anonymous. Beighton believed that mathematics disciplined the mind, a widely held opinion at this time. He wrote in the 1715 edition of the almanac that "There is no science in the world that does improve the mind of man so much as mathematics."[17] Acting upon this assumption, he advocated the importance of mathematics and increased the complexity and mathematical scope of the *Diary*'s problems. While John Beighton did not maintain the fawning solicitation of female readers initiated by his predecessor, he did, in principle, continue to devote the *Diary* to the entertainment and instruction of women. He was also actively assisted in his editorial work by his wife and another female relative, perhaps his daughter. Thus the *Diary* now had some direct female editorial input. In the 1718 edition of the almanac, Beighton acknowledged and proudly boasted of the *Diary*'s ladies' interest and accomplishments:

> ...that the rest of the fair sex may be encourag'd to attempt mathematicks and philosophical knowledge, they see here, that their sex have as

clear judgements, a sprightly quick wit, a penetrating genius, and as discerning and sagacious faculties as ours, and to my knowledge do, and
can carry them thro' the most difficult problems. I have seen them solve,
and am fully convinc'd their works in *The Ladies' Diary* are their own solutions and compositions. This we may glory in as the Amazons of our
nation; and foreigners would be amaz'd when I shew them no less than 4
or 5 hundred several letters from so many several women, with solutions
geometrical, arithmetical, algebraical, astronomical and philosophical.[18]

These statements are as much a confession of personal discovery on the part
of the editor as a proclamation of feminine mathematical achievement and potential. The "philosophical" referred to is the subject of physics, primarily statics
and dynamics. While some modern feminist reviewers of these comments have
taken offense with the designation "Amazons", I believe that within the context
of the times it was a compliment, quite simply noting superior women.

Henry Beighton was a respected surveyor, draughtsman, cartographer and
engineer; so scientifically versatile that the author Alan Cook has described him
as "the Universal Hanoverian Man." Working as an engineer, Beighton made
a study of the then revolutionary Newcomen steam engine that had been introduced into England in 1712. This new, more efficient engine was employed in the
pumping of water from deep coalmines thus allowing for increased production.
In 1717 Beighton published an illustration of this mechanical marvel.[19] See Figure 3.3. In a more scientific vein, he collected data on the pumping performance:
quantity of water extracted based on pump stroke length and cycles per minute,
and improved designed parts for the engine. He included his Newcomen steam
engine findings in the 1721 issue of the *Diary*.[20] In this same edition, he strongly
associated engineering with mathematics and, in a sense, with the rising industrial revolution. Beighton warned against those who pretend to be engineers but
do not know their mathematics and only guess at answers, preferring:

> ...he who has skill enough in Geometry, to reduce the Physco-
> Mechanical Part to Numbers, when Quantity of Weight or Motion is given,
> and the force designed to move it, can bring Forth all the Proportions, in
> Numerical Calculations, so it may Be almost impossible to Err.[21]

Henry Beighton was elected a member of the Royal Society in 1720 and would
go on to publish four articles in the Society's *Transactions*.[22]

Throughout the existence of the *Diary*, many contributors used pseudonyms
in identifying themselves. This phenomena will be examined later, however, the
owner of the *nom de plume* "Adrastea", assumed to be female, had consistently
distinguished herself both in enigma solving and mathematical computation. In
the 1720 *Diary*, she is recognized for having calculated the phases of the moon
as well as the dates and times for the year's eclipses. Perhaps chafing at this time

under his involvement in many tasks, Beighton wished to alleviate himself of the editorship and in the same issue offers an invitation/plea to Adrastea:

> [T]he calculations of the eclipses and limitations, are the actual perfor-
> mance and contribution of the fair-sex; who here merits my grateful ac-
> knowledgement. It was with difficulty, some years since, I discover'd her
> true name from that feigned one she assum'd but that stroke of modesty
> I cannot quarrel; at I confess, having not yet to the world discover'd my
> own. But having now a desire of her assistance, and being convinced of
> her ability and good-will to the design, perhaps the succeeding *Diaries*
> may come abroad under hers.[23]

With this statement, he is openly seeking a female editor. The offer was not accepted and Beighton continued as editor until his death. During his tenure he had moved the journal's thrust in a more serious mathematical direction and, while periodically reassuring and praising the ladies of their abilities, he also more openly solicited male interactions. Furthermore, he advocated the *Diary*'s mathematical learning benefits to schoolboys. Thus, he broadened the audience from that conceived by his predecessor. While nominally devoted to women, *The Ladies' Diary* was now openly intended for both sexes.

When Henry Beighton died suddenly in 1743, the Company of Stationers allowed his wife Elizabeth to complete the 1744 issue with "the assistance of a deputy." She was the actual editor, satisfying her husband's wish—*The Ladies' Diary* finally had a female editor, however, only briefly.[24] The exact identity of this assisting deputy remains confused, the most likely candidate seems to be Anthony Thacker, a mathematics master from nearby Birmingham, a mathematician of some talent, and an enthusiastic problem contributor and solver for the *Diary* since 1725. He was well known to the Beightons, assisting them at times with the compiling and editing of the problem section. Thacker had also served as a contributor/author to the almanac the *Gentleman's Diary or Mathematical Repository*, another of the Stationer's publications. He collected problems from both journals with the intention of publishing a three-volume series: *A Miscellany of Mathematical Problems*; however, his sudden death by fever in 1744 curtailed this project. Only one volume of his intended work appeared, published in 1743. It contained all the problems and their solution of *The Ladies' Diary* for the year 1744. Thacker would most likely have followed Henry Beighton as editor of the *Diary* had he lived. The other possible candidate was a Robert Heath, also known as a past contributor to the *Diary* who had a strong interest in the periodical.

Figure 3.3. Thomas Newcomen's 1712 atmospheric engine was the first fully practical, commercially successful steam engine and genesis for the industrial revolution. The full title of the engraving is "A description of the engine for raising water by fire." The image shown is from a newspaper article 1725, with Leyourn's original engraving altered by London engraver Nicholas Sutton. Source: Held by the British Library, Public Domain http://creativecommons.org/publicdomain/mark/1.0.

The Company of Stationers formally appointed Robert Heath (1720–1779) editor of *The Ladies' Diary* in 1745.[25] Heath was a military engineer by profession, recorded as "a captain on half pay" in the British army and had been a frequent contributor to *The Ladies' Diary* even supplying several prize questions for the journal: 1739, 1740, and 1742. He had actively assisted Henry Beighton with the compilation of mathematics questions; so he was well known to the Beightons. Elizabeth Beighton edited the 1744 issue. She remained involved with *The Ladies' Diary* but Robert Heath would be the designated editor.

Robert Heath was a contentious and self-serving individual, the central figure in several disputes that were soon to arise. In the parlance of the day, he would be described as a "rascal." In principle, *The Ladies' Diary* was still intended for the entertainment and enlightenment of the fair sex; however, in design and content, under Heath's editorship it was still becoming more serious as a mathematics journal, an obvious venue for mathematics problem solving. Editorial duties were divided between Elizabeth Beighton: she responded to correspondence and compiled the enigmas, while Heath focused on scientific entries and the mathematical problems. In general, the problems became more complex and difficult, beyond the capacity of the average amateur mathematician, male or female. Heath would eventually be accused of overburdening the *Diary* with mathematics problems that were of limited interest, and which would serve an estimated 500 readers at most.[26] Some such objectionable problems were submitted to the *Diary* by its editor including the Prize Problems of 1746 and 1748, the latter for which he awarded himself the prize for the correct answer. This question was:

> On what Day of the year does the City of London travel
>
> the greatest and least Number of Miles by the *diurnal* and
>
> *annual* Motion of the Earth? And how many miles per Day
>
> and also per Hour about Noon does it travel, when the Days
>
> are *longest* and *shortest* in this place?[27]

As a result of these changes and self-serving trends on the part of the editor, the *Diary*'s subscriptions decreased markedly. The contentious editor also constantly found fault with his employer, The Company of Stationers. He resented their monopoly on the printing of almanacs and felt his salary was too low. Heath openly complained of his situation.

In an attempt to attract more readers and stimulate sales, in 1749, Robert Heath announced an extended scheme for improving the *Diary*, the addition of a new feature, "The Ladies' Oracle or Querist." It was a question and answer column where correspondents sent in questions that would be answered by other correspondents the following year. Such a feature was common to many periodicals of the era and was particularly attractive to the ladies. The "Querist" addition was well received and sales rebounded.[28] The first question asked was submitted by a "Lady Nunquam Satis."

Heath composed and published instructional essays, sometimes employing his real name and on other occasions using a pseudonym. Such "cover-ups" were colorful and imaginative including: *Upnorensis, Newtoniensis, Critic Ansr* and *Hurlothundro*. Mathematically speaking, Newton's fluxions and the new calculus were the current center of scientific excitement and controversy. The concept of "the infinitesimal," or very small/instantaneous mathematical change, was found philosophically, and even morally, questionable by many scholars. Robert Heath was an active advocate and promoter of infinitesimals. In the 1746 *Diary*, he published "A New Account of the Nature and Idea of Fluxions"[29] intended

"...to annihilate all objections, doubts and absurdities" surrounding an understanding of fluxions. The piece was continued in the 1747 edition. In the 1751
and 1752 issues, he discussed "the Algebraic Quantity 0"[30], a relevant concept in
the understanding of infinitesimals. Using pseudonyms, Heath also contributed
two mathematical problems to the 1752 *Diary* of a rather risqué nature, full of
innuendo: problem XIII, set by "Honorius" concerns fitting a young lady with a
"French hoop" and computing the volume of the lady's petticoat and that encom-
passed by the worn hoop. The lady attracts attention by revealing her charms:

> Her hoop is the secret—and if you would know What it holds with her
> Petticoat, seek from below.

Furthermore, the required volume of petticoat enclosure was to be calculated in
wine gallons (an innuendo?); and the "Prize Problem", regarding a duplication of
Archimedes' bath episode to determine the volume of a mathematician with an
attached note (N.B.) that the experiment can be performed on a woman where
curves and volumes can be determined.[31] Such tantalizing problems would be
happily received and discussed by the "chaps" in the coffeehouses but found
scandalous and embarrassing by lady readers.

Seeking additional publishing opportunities and the chance for further financial gain, Heath conceived of and published a series of other periodicals: the
Palladium (1749), *The Ladies' Philosopher* (1752) and *The Ladies' Chronologer*
(1754). All were released without the approval of the Company of Stationers and
without payment of the required government stamp taxes.[32] To make matters
worse, the *Palladium*, promoted as an *Appendix to The Ladies' Diary*, siphoned
off materials intended for the *Diary*, to the benefit of Heath who openly promoted
his new journal in the *Diary* as in the 1749 issue:

> Hail happy Ladies! Take your joy complete,
> Your *Diary* and your lov'd *Palladium* greet.
> No tasteless art degrade your lively wit.
> Like hand and glove, so charmingly they fit.[33]

In the 1751 issue of the *Palladium*, Heath published a solution for the "Prize
Question," 22 of the 1750 edition, submitted by a one "Sophia Western," knowing that the solution was incorrect. He had also set the question and answered
it himself under the name "Newtoniensis." The problem required a use of fluxions and fluents (calculus). Heath then went on to berate her on her mistakes and
noted "...yet we would not deprive her of publick judgment" [humiliation?]. Perhaps some correspondents would welcome such interactions, others, less secure
in their mathematical abilities, would be deterred from submitting their solutions
to any journal associated with Robert Heath.

With the exception of the *Palladium* which ran for thirty years until 1778, the
journals were short lived. The *Palladium* marketed by Heath at a lower price than

The Ladies' Diary became a serious competitor to the Stationers' offering. Further adding to a climate of discord was Heath's apparent trait of carrying a grudge and belittling people who did not always agree with him. In 1736, he approached Thomas Simpson (1710–1761), then schoolmaster in Nuneaton, Warwickshire, asking to be tutored in mathematics. Simpson, known as an exceptional teacher and extraordinary mathematician, refused Heath's request for instruction, apparently setting the man against him. In 1737 when Thomas Simpson published *A New Treatise on Fluxions* [**Sim37**], Heath railed against the book's treatment of fluxions. The editor of *The Ladies' Diary* even carried his condemnations of Simpson's work into his fluxion essays of 1750 and 1751 and also into the *Monthly Review* of 1750, employing the pseudonym "Cantabrigienis."[34] In contrast to this condemnation of Simpson's treatment of fluxions, he constantly praised and promoted the similar work of William Emerson (1701–1782), especially his *The Doctrine of Fluxions*, 1743 [**Eme43**].[35] Included among Heath's denunciations was an added charge of plagiarism against Simpson for imposing upon *Estimato Errorum*, a work of Roger Cotes (1682–1716).[36] John Turner, a former student of Simpson's, and an able mathematician in his own right, responded to Heath's condemnations in the 1751 issue of *Mathematical Exercises*, defending his teacher's work and, in turn, questioning Heath's ability and opinions; furthermore, he called attention to Heath's purloining of the *Diary*'s materials in the *Palladium*.[37]

By this time, The Company of Stationers had endured enough of Robert Heath's offensive behavior and transgressions. They fired him. And, seemingly in ironic vindication, hired Thomas Simpson in 1754 as the new editor for *The Ladies' Diary*. In the 1754 edition of his journal *The Ladies' Chronologer*, Heath claimed that he left the *Diary* because of "ill treatment" and attacked his adversaries in a dialogue involving "Tom Tickle of Pickle Hall" who recites a little ditty:

> Bid Turner ignorant derider
>
> Mind how to bring his master cider.
>
> Simpson to Bosworth did repair
>
> There weave his cobwebs in the air
>
> There propagate fictitious lies,
>
> And calculate nativities.[38]

In later years, the distinguished British mathematician and mathematics educator, Augustus De Morgan, would condemn Robert Heath as "a person who made noise in his day" and a "worthless vagabond" and noted his use of indecent double-meaning in the composition of mathematics problems.[39]

These three individuals, the first editors, established the basic format, an almanac of three major parts including: a reference calendar, mathematical problems and word problems or enigmas, and the problem sections would be reader/correspondent driven and interactive. In total, their actions also set a trajectory for the periodical. John Tipper designed and intended a publication specifically for ladies. He was sincere in his goals, acting in accord: fawningly encouraging his female readers and censoring the difficulty of the mathematics to their perceived, dictated abilities. Henry Beighton remained more remote of the ladies than his predecessor, not overly reaching out to his female readers. He "opened a door", inviting male participation in the *Diary*'s problem-solving activities and increased both the scope and difficulty of the mathematical problems. Robert Heath's temperament, actions, and reputation alienated ladies, lessening the number of female subscribers. His personally set problem involvement in proposing obscure and even ridiculous solution constraints deterred solvers while his ribald innuendos embarrassed and shocked proper ladies who would remove themselves from such taint.

Thus *The Ladies' Diary* within its first half century of existence moved from being a primarily female-intended journal to one available to the general public, male and female, and whose mathematical problem challenges morphed from the "genteel" to the '"more worldly." These trends would increase with the following editors.

3.3 Sustainers of the *Diary* and Advocates of its Mathematics

Through Heath's editorial mismanagement, the *Diary* had suffered badly, its reputation tarnished as taunted in a contemporary rhyme:

> *The Ladies' Diary* once known to
> Fame
> Now casts dishonor on the lady's name![40]

The new editor of 1754, Thomas Simpson, assisted by Elizabeth Beighton now worked to restore the *Diary* to its previous, pre-Heath standing. Simpson, a self-trained mathematician and rising star on England's mathematical horizon, brought a variety of experiences to his editorship.[41] At his appointment, he was already a faculty member at the Royal Military Academy (RMA) at Woolwich, a training school for British artillery officers, and a Fellow of the Royal Society (FRS), making him the most highly qualified and mathematically active person to occupy the position up until that time. With his editorship, the level of mathematical presentations improved greatly and a new form of puzzle, the "Rebus" was introduced in the 1754 almanac.[42] Despite his editorial chores, Simpson was quite occupied as an author, publishing: *The Doctrine and Application of Fluxions*, 1750; *Select Exercises for Young Proficients in Mathematicks*, 1752, and *Miscellaneous Tracts*, 1757 as well as having several of his papers appearing in the

Philosophical Tracts of the Royal Society.[43] Due to failing health, he resigned as editor of *The Ladies' Diary* in 1760, having produced only six issues of the *Diary*.

The next editor selected by the Company of Stationers was Edward Rollinson (d. 1773). Apparently, he was an able and willing Philomath, having solved several problems published in both the *Diary* and the short-lived journal the *Mathematician* (1745–1750) for which he also served as editor. Rollinson was a diligent caretaker/editor of *The Ladies' Diary*, he neither innovated nor detracted from the journal as restored to respectability by Simpson. After overseeing the publication of twelve editions of the *Diary*, his editorship was cut short by his death in 1773.

The Stationers appointed Charles Hutton (1737–1823) editor of the *Diary* in 1774 replacing the deceased Rollinson. Hutton was a well-respected mathematician and educator. A surveyor by profession, mainly self-educated in mathematics, he had risen from the coalfields of Newcastle to become a schoolmaster and founded his own school. As Professor of Mathematics at the Royal Military Academy at Woolwich, he was the most experienced mathematician to occupy the editor's position at *The Ladies' Diary*.[44] See Figure 3.4. Well known as a teacher, applied mathematician, and textbook writer, his reputation would help embellish the status of the *Diary* and increase the almanacs' audience. Furthermore, he had contact with a network of mathematically talented people within England: students, teachers and his colleagues at the Military Academy. Many of these "Philomaths" would now contribute to the *Diary*.

Hutton would also be the longest serving editor, ending his term in 1818 after overseeing the compiling and publication of forty-four issues of the almanac. Seventeen seventy-four also marked the year Hutton was elected to the Royal Society. Under the Society's jurisdiction, his major scientific contribution was a calculation of the mass and density of the Earth.

Even before his editorship, Hutton had been an ardent contributor to the *Diary*, proposing and solving problems: in the period 1764 to 1773, he contributed twenty-four questions and answers. Impressed by its contents, he had begun in 1771 collecting "all the useful and entertaining parts, both mathematical and poetical" from the almanac since its inception in 1704. He started to publish this collection serially under the title *The Diarian Miscellany* but in 1775 combined all this material into a five-volume work bearing the same title with a scope of coverage from 1704 through 1773. In a very similar manner *The Darian Repository or Mathematical Register*, was supposedly compiled "by a society of mathematicians" but apparently its suspected editor was Henry Clarke, a mathematician and the son of Samuel Clarke. This periodical claiming to contain all the *Diary*'s mathematical problems and their answers from 1704 onwards began its publication in 1771. The *Repository* was conceived "As an Easy and Familiar Praxis for Young Students in Mathematical and Philosophical Learning". It existed for four years, ceasing publication in 1774, and covered the *Diary*'s mathematics up to 1760. Hutton also added a "Supplement" to the *Diary* that was published in conjunction with the *Repository* from 1788 to 1800. In 1786, a new puzzling feature, "Charades" was also included in the *Diary*.[45] Due to ill health, Charles Hutton resigned from his Woolwich professorship in 1807 but still continued on as editor of *The Ladies' Diary* until 1818.

Figure 3.4. Contemporary engraving of influential mathematician and *Diary* editor Charles Hutton (1737–1823) by William G. Jackman. Source: Wikimedia Commons.

The next editor, following Hutton, Olinthus Gregory (1774–1841), received his early mathematical training at the boarding school of the recognized botanist and mathematician Richard Weston. The boy's scholarly talents soon became evident. Encouraged by Weston, he began to solve mathematics problems posted in *The Ladies' Diary*. At the age of nineteen, he published his first book, *A Treatise on Astronomy* (1802). Gregory and his abilities came to the attention of Charles Hutton who arranged for the young man to be appointed a second mathematics master at his Royal Military Academy (RMA).[46] Besides his teaching duties at Woolwich, Gregory became a close associate and friend of Hutton, supporting him in several tasks including his editing of the *Diary*. Thus, when Hutton stepped down as the *Diary*'s editor in 1818, it was appropriate that Olinthus Gregory assume the editorship. He officially became editor in 1819 and would continue in the post until 1840. In 1821, he rose to the rank of full Professor at the Academy. When in 1835 the Stamp Act was repealed reducing the production costs of the *Diary*, Gregory used this opportunity to increase the size of his periodical by adding appendices in which expository essays on topics in mathematics and science would be added.[47]

In 1841, *The Ladies' Diary* was amalgamated with the *Gentleman's Diary or Mathematical Repository*, another periodical owned by the Company of Stationers. This combined effort became *The Lady's and Gentleman's Diary*, "designed principally for the amusement and instruction of students in mathematics." See Figure 3.5. Now more openly a periodical of recreational mathematics, this descendent of *The Ladies' Diary* would continue until 1871. The new journal's major editor (1844–1865) was Wesley S.B. Woolhouse, a mathematician and instructor at the Royal Military College, Sandhurst.

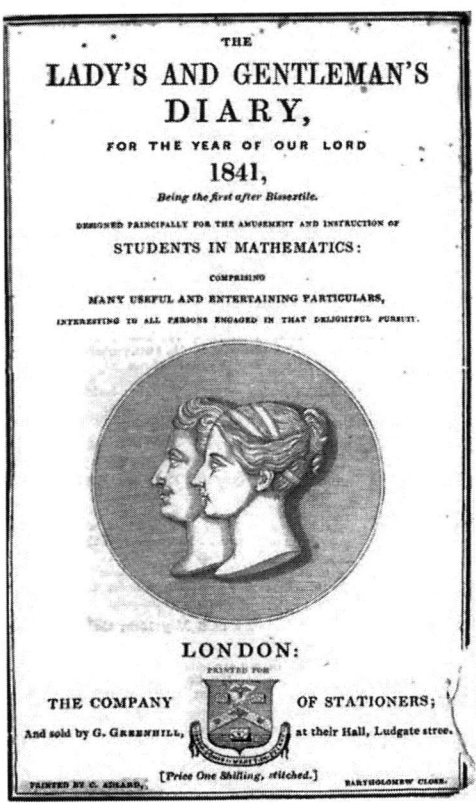

Figure 3.5. The cover of *The Lady's and Gentleman's Diary*, 1841. This successor to *The Ladies' Diary* was "designed principally for the amusement and instruction of students in mathematics." Courtesy of Wikimedia. The issue cover shown is held by Harvard University. Rights: Public Domain, Google-digitized.

Thus from 1754 to 1840, all the *Diary*'s editors were mathematicians and former problem solvers for the periodical. Most were associated with military academies. They stressed their academic discipline and interests, with the result that mathematical problems became more complex and difficult, requiring specific advanced knowledge. Ladies denied the appropriate educational training were seriously limited in participating in such problems. Mathematical solutions posted by women problem solvers became almost nonexistent and the feminine focus and initial intent of *The Ladies' Diary* became diminished. While over the course of changing editorship some aspects of its original design diminished, others, especially a concern with mathematics, were added. These transitions converted the *Diary* to a journal that focused on problem solving.

4

"Delightful and Entertaining Particulars" — Problem Solving

Why enigmas and mathematical problems?

4.1 Some Thoughts

Within the past thirty years, much educational attention has been focused on the topic of problem solving, specifically answering the questions of: "How to pose good problems?" and "How to teach the skills of problem solving to children?" A problem exists when an individual or organization is confronted with an obstacle to be overcome. Such obstacles come in many forms: physical, intellectual, psychological and in many other guises. Human existence, and indeed survival, can be described as a sequence of problem-solving events. In school, problem solving is most easily associated with mathematics although learning in any subject involves finding the solution to problems.

In a mathematics class, a student is given a problem, perhaps in symbols such as numbers or x's and y's, or in words, and asked to find an answer; hopefully, the correct one. A process for doing this has been instilled: "What is asked for?"; "What is given?"; "Have you done any problems like this before?" ... The student should be cognitively engaged, see a direction in which to proceed, a strategy for obtaining a solution, and if that does not get the desired results, alter the strategy or attempt another. She should not become frustrated.

One overriding factor in this scenario is that the one confronted by the problem, the assumed problem solver, <u>must want to solve it</u>. The problem must be attractive in some sense-relevant, interesting, challenging or, at best, a combination of these qualities.

The Ladies' Diary was a periodical devoted to problem solving. This feature, combined with the journal's success in longevity, reflects directly on its audience and the changes taking place within the social milieu. Certainly, enigmas, paradoxes, rebuses charades, and mathematical questions are all forms of problems. Since the two features attracting most participation were the enigmas and the mathematical questions, they will now be analyzed in more detail as forms of problem solving.

4.2 The Enigma, A Word Maze

A modern reader would probably ask "What's with these enigmas?" And rightly so, for most of us do not encounter enigmas in our daily lives, or if so, they are in forms we don't recognize. For a Lady or Gentleman in eighteenth- and nineteenth-century England, the situation was very different. Such word puzzles were the fashion, an occasion to demonstrate one's grasp and use of the language, to be "witty" without being offensive, and to demonstrate a cognitive ability to unravel and solve problems. Simply, to show how smart you were.[1] John Tipper was well aware of this fashion.

An enigma, in current parlance, is something that is difficult. It can be a person, a situation, or a statement that requires unraveling and understanding. The lyrics of contemporary "rap music" songs, appear as enigmas to the uninitiated. For the readers of *The Ladies' Diary*, their enigmas were written, and often answered, in verse, another fashion of the time. Verse was more genteel, easy on the ear and befitting Ladies and Gentlemen. The presence and solving of enigmas did not begin in Hanoverian England but are thousands of years old reaching back into Ancient Babylonia and Classical Greece. Enigmas can be found in all ancient languages such as Hebrew, Sanskrit, and Chinese. Professional riddlers amused nobles in Imperial Rome.[2] Enigmas have always been a part of human communication and interaction. They impart a sense of power to the poser—"I know something you don't and you must find it out." Some contemporary readers may remember playing "Twenty Questions" as a young partygoer; a game with similar objectives. Lexical and grammatical ambiguities must be overcome. Riddles are a test of knowledge and a nonthreatening challenge, many times evoking laughter among their audience. In their composition and solution, they usually tell something about the culture and society in which they were conceived.

Let us examine and analyze two enigmas from the 1835 *Diary*. The number referencing serves two purposes: the Roman numeral specifies the order of the enigma within the particular issue of the almanac; the Hindu-Arabic numeral designates the position of the enigma within the total collection of submitted enigmas.[3] These are among ten such word riddles in this issue that had been sent in by *Diary* correspondents:

III. ENIGMA (1173); *by* Mr. J. OXENFORD, 37, *Finsbury Circus.*

The regal crown shines bright with many a gem,
But I oft bear a brighter diadem;
I own indeed no purple robes I wear, [more fair.
But yet the lily's self is scarce
A friend, like great Medea, in dark times, [climes.
Am I to learning, patron of all
E'en Jove, when he was ruler of the sky, [I ;
Never received more sacrifice than
The lamb, which on the mead is sporting free, [me.
Must soon be a burnt offering to

The mighty whale, the sov'reign of the main, [slain.
For me is captured, and for me is
The bee may build his cells with instinct fine,
Those waxen homes must melt before my shrine.
And when the sun has fallen from the skies,
Then, clad in all my glory, I arise;
I scorn the day, but glory in the night,
And only shine when I alone am bright.

IV. ENIGMA (1174); *by* Mr. J. OXENFORD.

X. *or* PRIZE ENIGMA (1180); *by* Mr. W. LESTER, *Woodhouse.*

[Whoever answers it before Feb. 1, has *two* chances for *twelve* Diaries.]

The Muse long us'd to bask in Dia.'s smiles, [wiles ;
Again essays t' amuse with mystic
Discard not then, your humble servant long,
Oblivious, like the subject of my song.
Form'd for your ease in ages long since past, [cast ;
Your cares and pleasures on my bosom
To you alone my service was devote,
In humble guise, bedeck'd in woollen coat :
By day or night a stedfast willing slave, [outbrave ;
For you did floods and midnight storms
Pleas'd with my charge, so faithful did I serve,
Proud man ne'er made me from my duty swerve.

When great Eliza ruled the rising realm,
Wisdom and valour guiding at the helm,
The infant arts fresh lustre did obtain,
Conspicuous glories of a female reign :
She knew me well, nor would my worth disown,
And frequent made me her imperial throne ;
Bent at my footstool—then I nobly great,
Upheld the burden of her lofty state.
A favorite long with ladies it appears,
Support of weakness and declining years ;
Females of every rank and every grade;
My help disdain'd not, widow, wife, or maid :

PRINTED FOR THE COMPANY OF STATIONERS.

Both enigmas contain references to Classical themes and beings, a popular subject of interest at this time, as were biblical events. As beginners at the task of unraveling such quandaries, let us attempt to glean clues for an answer for Enigma III. Remembering that these are composed for a late eighteenth-century audience, word play is a paramount feature. Some terms and expressions must be interpreted within this contemporary milieu. After reading the puzzle several times, certain phrases and events seem to stand out:

"... regal crown stands bright" } → bright (light?) → Function

"... bear a brighter diadem"

"... received more sacrifice than the lamb"

"the mighty whale, the sov'rerign

of the main,

For me is captured and for me is slain." }→ fat, oil, → Substance

"The bee may built his cells with   wax

of instinct fine,

Those waxen homes must melt before my shrine."

"... the sun has fallen from the skies

Then clad in glory I arise }→ used at night → Purpose

Then:

(Function & Substance & Purpose) → candle

Answer, candle."

The particular clues are isolated and a solution is obtained through a logical chain of deductive reasoning. A process of deductive problem solving is employed which in its application and structure is the same process used to solve mathematical problems. Heath attempted to convince readers of this fact—that mathematics problems were also enigmas. The prize enigma, X, will be left for the reader to ponder further with one assisting hint: "for your ease" = comfort. (The reference to *Dia* in the first line is the affectionate term that the readers use for the *Diary*, i.e., *The Ladies' Diary*.)[4]

4.3 The Enigma's Enduring Popularity

Early historical evidence reinforces the hypothesis that mathematical problem solving and enigmas are closely related. One of the earliest recognized collection of mathematics problems *Propositiones ad acunendos juvenes* (*Problems to Sharpen the Young*), a collection of 56 mathematical puzzles, was devised in the eighth century.[5] Originally intended for the education of youths in Charlemagne's court, many of these puzzles have appeared over the ensuing centuries in various lists of mathematical exercises and textbooks. Perhaps the most famous of these, often clothed in a variety of cultural and societal guises, concerns a river crossing:

> A man had to take a wolf, a goat, and a bunch of cabbages across a river. The only available boat he could secure could take only two at one time. But he had to transport all of these to the other side in good condition. How could he do it?[6]

The problem's author, Alcuin of York (ca. 732–804), so liked this challenging situation that he employed it in two other of his puzzles: one involving three virgins and their respective brothers who were possible seducers of the women in this party other than their sisters and a family of husband and wife with children, whose combined weights hindered the transport arrangements. Several, later, notable mathematicians used and studied this "river crossing problem", among whom are: Luca Pacioli (13th century), Tartaglia [Nicolo Fontana] (1556), Gaspar Bachet de Meziriac (1612), and M. Cadet De Fontenay (1879). By the twentieth century, this problem situation and its implications became a subject for mathematical research as "transport" and "network analysis" problems.

Riddles and enigmas could always be found in literature. Geoffrey Chaucer (1342–1400) in his *Canterbury Tales* perplexes his reader with the situation of four competing pilgrims: John, Geoffrey, Martin, and Stephen, on the road to Canterbury, boasting of their forthcoming benevolence upon arrival at their destination.

John says: "I usually only tithe two pence, but if I beat Stephen to Canterbury, I'll gladly tithe double!"

Martin utters: "If I get to Canterbury first, I shall show my gratitude by tithing six pence! If I don't, I shall tithe four pence anyway."

Stephen responds: "You wayward fools are tossed by the wind! No matter what happens before here and Canterbury, I shall tithe three pence only. Heaven forgive your false piety!"

Geoffrey speaks: "Shut up Stephen. I pledge three pence if I beat you to Canterbury, but nothing if I don't. We'll see who's right!"

Everyone was true to his word, and a total of thirteen pence was tithed at Canterbury.

The question remains "In what order did these pilgrims arrive?"[7]

Jane Austen (1775–1817), the British novelist whose books and characters have contributed much to an understanding of the life and customs of the landed gentry of the period we are investigating, was noted for her riddling ability and capacity. She incorporated several riddles into her novels.[8]

4.4 The Enigma in the Period of *The Ladies' Diary*

So throughout history, solving enigmas and finding the answers to mathematical questions have always been closely associated as forms of problem solving. But "Why did they so appeal to the British audience of the eighteenth and nineteenth centuries?" First, one of the benefits of economic prosperity is the rise of a leisure society. The Ladies and Gentlemen were members of this leisure society—they required pastimes and entertainment. Enigmas helped fill time and favorably reflected on the intellect and wits of their proponents. At that time parlor games were very much a part of the social scene where the interaction between men and women via an enigma posing/contest would not be considered improper. Ladies actually carried around small notebooks in which they could record new enigmas. The intellectual dynamics also contained the psychological factors previously mentioned. Across the channel, the French had their fashionable "witty" salons; the enigma interchange of *The Ladies' Diary* served a similar function but on a more egalitarian scale.

The long life of *The Ladies' Diary* has been attributed to its enigma feature. This is true, but also true is that throughout the *Diary*'s existence, Ladies excelled in the composing and solving of enigmas. They quite amply and continually demonstrated their ability as formal problem solvers where they worked from a set of premises and logically arrived at a solution! Now Dear Reader, we leave this section with one more of the *Diary*'s wordy gems from the year 1744 for you to unravel:

> III Enigma 262, by Miss Ch__bers.
>
> I am a very useful thing, extracted from the earth
>
> By art and labour, roughly us'd before and after birth.
>
> My maker's ingenuity appoints the shape I wear.
>
> Sometimes like a wheel am round but mostly I am square.
>
> Tho' homely be my garb and mien, in courts of kings I'm us'd
>
> Lord O__d he made use of me, or else he is abus'd.
>
> In almost every family I'm held in great request.
>
> Because I'm known to give new gust to scraps of Christmas feast.
>
> Further I say, and true I may, that altho' I am able
>
> To fill the purse and belly too, I ne're appear at table.
>
> Now, ladies, as I'm pretty sure, each of you is a lover
>
> Of what I prepare for you, I pray my name discover.[9]

For the more adventurous wordplay enthusiasts, several more challenging word puzzles from *The Ladies' Diary* can be found in Appendix A.

4.5 The Mathematical Questions

A few selected mathematical questions will be examined here and relevant implications offered but the mathematical impact of *The Ladies' Diary*'s mathematical questions is so significant that it deserves further attention to be given below in a separate chapter. Let us now look at two categories of these questions: questions that are already referenced in passing, previously having been mentioned above in other contexts, and a few questions representative of each, individual editorship, noting how their emphasis and tenor changed.

Here are the first two mathematical questions to appear in *The Ladies' Diary* (1707). They were offered by a Mr. John White of Butterly, Devonshire. The copies of problems shown here with their answers are from a later collection assembled by Thomas Leybourn in 1817:[10]

1. QUESTION 1.

In how long time would a million of millions of money be in counting, supposing one hundred pounds to be counted every minute without intermission, and the year to consist of 365 days, 5 hours, 45 minutes?

Answer. 19013 years, 144 days, 5 hours, 55 minutes.

Solution.

The solution of this question is evidently thus : As 100l. : 1 minute :: 1000000000000l. : 10000000000 minutes = 19013 years 144 days 5 hours 55 minutes, the true time required. H*.

This is a simple computational problem whose impact is the amount of money being considered. In the 1709 *Diary*, the originator Mr. White relates a discussion he heard by local ploughmen (farmers) concerning the problem's solution. In reporting this conversation, he attempted to capture the tone and dialect of these men in a verse:

Says *Tom* 'twol be vorty long Days, —— 40 days?

I and vorty to that says *Will*:———— 40 + 40 = 80 days?

'Twant be told in a Year quoth *Jack*,—— a year?

No nor in zov'n Years cries *Jill*:———- not even 7 years?

You talk all like Vools saith *Roger*.

A *Merchant* with's two vore Veet,

Will scrape it away in a Month,———— a month?

And thereto I'll wage you a Sheep.

Go Blockhead quoth *Bess* that was brewing,

The *Boy* that weighed my Hops,

Woll tell it all in a Week,   } —— a week

Zo will any Mon in the Shop[11]

Obviously, these men are just guessing but this bucolic conversational scenario attests to the wide spread readership of *The Ladies' Diary*—even farmers were reading it or at least knew of and were discussing its questions!

II.   QUESTION 2*.

If to my age there added be
One half, one third, and three times three;
Six score and ten the sum you'ld see,
Pray find out what my age may be.

Answer.   66 years.

Solution.

The meaning of the problem is, that the number 9 added to once his age, together with one half and one third of his age, the sum shall be 130; or since the sum of the parts 1, and $\frac{1}{2}$, and $\frac{1}{3}$ is $\frac{11}{6}$, that $\frac{11}{6}$ of his age is $(130-9=)$ 121; consequently $11:6::121:66=$ his age.   H.

Algebraic Solution.

Let $6x$ represent the required age; then, by the question, $6x + 3x + 2x + 9$, that is $11x + 9 = 130$; therefore $x = 11$; consequently $6x = 66$, as before.   L.

---

* All the solutions marked with the signature H, are by Dr. Hutton, and taken, with permission, from his edition of the Ladies' Diaries.

This, II, is a simple "find my age problem", a type of puzzle used for over a thousand years and one that would often appear in the *Diary* in various forms.[12]

A lady, Mrs. Mary Wright, answered the first-prize mathematical question, (1710), VI, 16. During the early years of the *Diary*, Mrs. Wright was a consistent and fervent mathematics problem solver and composer. She demonstrated her ability by eventually solving several "prize questions". To obtain her answer for the steeple problem, she had to have had access to longitudinal reference tables. Mary often joined with her nearby cousin, Thomas Wright, in solving problems. Such informal problem-solving gatherings, people working together, finding solutions to the *Diary*'s mathematical questions, were not uncommon. The *Diary*'s exercises encouraged these interactions within families and groups of like-minded friends. The question Mrs. Wright confronted, as summarized by Leybourn, and her answer are given below:[13]

VI. PRIZE QUESTION 16.

Walking through Cheapside, London, on the first day of May, 1709, the sun shining brightly, I was desirous to know the height of Bow steeple. I accordingly measured its shadow just as the clock was striking twelve, and found its length to be $253\frac{1}{8}$ feet ; it is required from thence to find the steeple's height.

Answered by Mrs. Mary Wright.

May 1, 1709.

	°	′	″
Sun's longitude, from its ingress into aries	51	28	0
Oblique angle of the ecliptic and equator	23	29	0
Thence the declination that day	18	9	45
Consequently its merid. altitude in lat. 51° 32′	56	37	45
The complement thereof to 90 is	33	22	15

Then as the sine of the angle 33° 22′ 15″.
To the base 253·125 feet.
So is the sine of the angle 56° 37′ 45″.
To the perpendicular 384·307 feet the height of the steeple.

Note. The true height of Bow steeple is 225 feet, for which at first I had proportioned the length of the shadow, but upon second thoughts I altered it, for fear some, who had read its height in history, should claim the reward, without having art enough to investigate it by trigonometry.

The mathematics involved in obtaining the solution is not difficult but required finding the sun's angle of inclination for Cheapside, at noon on May first, 1709.

The following "ribald" question was composed by Robert Heath, under the alias "Honorius", and offered in the 1752 *Diary*.[14] This question ultimately lost him the editorship. Miss Polly's French hoop is probably the latest fashion craze acquired from Paris. Hoops were fashionable and carried to such extreme dimensions as to force a wearer to struggle sideways through door openings. Although, in general, the English expressed a poor opinion of the French—being traditional enemies—yet French manners, fashions, and social behavior were much admired and imitated in British society at this time. See Figure 4.1

Figure 4.1. French-inspired women's attire was elaborate but not very practical. Several magazines were founded to serve British ladies' needs, primary among which was information on the latest French fashions. One such magazine was *The Ladies' Magazine*, 1770–1847. This image is entitled "A Lady in Full Dress in Augt. 1770" and was published in *The Lady's Magazine*, August 1770. Source: Wikimedia Commons.

Just as Editor Heath used a pseudonym in posing the problem, so too did the respondents: Messrs. "Honey" and "Wigglesworth." The abundant use of pseudonyms in correspondence to the *Diary*, both in posing puzzles and questions and submitting answers, can easily be attributed to the protection of one's identity and social standing. Contributors did not want to look "foolish" or ignorant, nor write anything that would diminish their position in society. They wished to protect themselves, remain anonymous, and yet have some pseudo

XIII.   QUESTION 359, *by* Honorius.

Miss's apron grown short, she is full of complaint,
And to merit your pity she looks like a saint!
On the floor falls her tea; then her screams you may hear,
And fainting she sinks in a fit on the chair.
Mamma for the doctor immediately sends,
Who, in honour to miss, in his chariot attends;
He examines her pulse, and appearing so wise,
Descants on the languishing looks of her eyes—
But alas! neither spirits, nor letting Miss blood,
Specifics, nor preaching, are found to do good :
For a surgeon came in, who the cause did declare,
And the doctor's finesse, and his art made appear.
 Mamma now was told Miss's hoop was too small,
Therein lay her grievance, disorder, and all;
The question was ask'd—Polly sighing reply'd,
A French hoop will cure me, and so will a bride.
A hoop of the fashion to cure her disease,
Extends from her centre quite round to her knees :
In the right and left wing a French placket * is made,
To her elbows advancing, and forms a parade.

* *Opens and shuts, forms a pair of bellows, and rises and falls by
the means of strings or bowlings.*

E 3

identity. But there is another reason for this practice, a social fad, a quaint custom of the time. Be witty. You are doing puzzles—"Why not make your identity a puzzle also?"[15] The great variety of picturesque, often double entendre, names used in the *Diary* reflect on the creative wit of their owners: "Miranda Tell Truth", "Nelly Needless" and "Simon Cucko Esq." Ladies mindful of their reputations would find such a ruse convenient.

70   LADIES' DIARY.   HEATH.   ⌈1752-53.

Miss Polly to church now, or play can repair,
And wherever she goes is admir'd for her air!
At the sight of a beau, how her heart beats alarms!
While the winds swell her pride and her legs tell their charms :
Her hidden perfections she knows will invite,
Or ensnare the beholder, should chance give them sight.
    By the pow'r of her hoop Polly steps into fame,
By out-priding the rest she conceals her own shame ;
In the country she reigns o'er the 'squire and the clown,
O'er the lords and the fops she's triumphant in town.
Her hoop is the secret—and if you would know ⁓
What it holds with her petticoat, seek from below †.

† *Form of the hoop is the lower frustum of an ellipsoid, with its vertex next the head.*

Transverse ⎰ *diams.* ⎰ 42 inches ⎰ *above,* ⎰ 48 inches ⎰ *below.*
Conjugate ⎱   ⎱ 26   ⎱   ⎱ 29   ⎱

Altitude of the frustum 12 inches.

From the lower part of the hoop's circumference to the bottom of the petticoat, the form is an elliptic cylinder, by the petticoat hanging nearly perpendicular from thence: the altitude of which elliptical cylinder is 18 inches : Quere the content of the whole concavity in wine gallons ?

Answered by Mr. John Honey, of Redruth, Cornwall.

Put a and $b = 42$ and 26 inches, the transverse and conjugate diameters of the hoop above ; and c and $d = 48$ and 29, the dimensions of those below ; also $m = 12$, the frustum of the ellipsoid's altitude ; and $n = 18$ inches, the elliptical cylinder's altitude : Then, by a known theorem, $(ab + cd + \sqrt{(abcd)})m \div 882.36 = 50.54$ wine gallons, the content of the hoop's concavity ; and $cdn \div 294.12 = 85.18$ wine gallons, the content of the cylindrical concavity ; whence the concavity of both = 135.72 wine gallons.

Mr. *John Wigglesworth* answers it thus : Let $a = 42$ inches, $b = 26$ = transverse and conjugate diameters above ; $t = 48$, $c = 29$ = transverse and conjugate diameters below ; $h = 12$, $p = 18$, and $m = .2618$, then the content of the whole concavity $= (mh (ct + \frac{1}{2}bt + ab + \frac{1}{2}ac) + 3mtcp) \div 231 = 135.7416$ wine gallons.

Question "377" is the problem our young gentlemen at the Black Turk Coffee House in London were discussing and to which Mr. Elsmont Potter responded. Several other readers also correctly answered the question but Potter pushed farther requesting the author's age.[16]

II.   QUESTION 377, *by Miss* Maria A—t—s—n.

There are three cities, A, B, and C, lying in the same road; whereof the first is 136 miles distant from the second, and the second 104 miles distant from the third: From A to B a courier travelled in two days; and from B to C in two days more, diminishing his distance every day alike, from the first to the last. What number of miles did he travel each particular day?

Answered by Mr. Samuel Kolt.

Let $2a = 136$, $2b = 104$, and $2x =$ the common difference of each day's journey; then $a + x$, $a - x$, $b + x$, and $b - x$ will be the respective distances travelled each day: But the first + the third $=$ twice the second, that is $a + b + 2x = 2a - 2x$; whence $4x = a - b$, and $x = \frac{1}{4}(a - b) = 4$: Therefore 72, 64, 56, and 48, are the four distances required.

The same answered by Mr. W. Gawthorpe.

Put $x =$ the first day's journey, and $y =$ the common difference: Then $2x - y = 136$, and $2x - 5y = 104$ (per quest.); and by subtracting the latter equation from the former $4y = 32$: Whence $y = 8$, and $x = 72$.

Atkinson's retort to the request of her age from "Mr. Potter:"[17]

New MATHEMATICAL QUESTIONS *to be answered in the next Year's* DIARY.

I. QUESTION 391, *by Miss* Maria Atkinson.
(Addressed to Mr. E. P. *who took the Liberty to ask her Age.)*
FIVE times Seven and Seven times Three
   Add to my Age; the Sum will be
As many above Six Nines and Four
As Twice my Years exceed a Score:
From hence, *Sweet Sir,* my Age explore.

In this period of confining, imposed, social limitations that included male-female constraints, "correct courting" rituals, such written repartee often took the place of flirting. Within the pages of *The Ladies' Diary,* young people sometimes found members of the opposite sex whose expressed opinions matched their own or which they found attractive and sought to pursue the source further. This practice was a nineteenth-century forerunner of online dating. For example, in a 1758 *Diary* encounter, the male inquirer, a "Mr. U. T__r", is more brazen, requesting a more detailed description of a female correspondent's status including her worth (possible dowry). Her response and its answer were published in 1759, whether the gentleman found the information sufficiently attractive to pursue the correspondence farther, we shall never know.[18]

I.   QUESTION 434, *by Miss* T. S—e.

Addressed to Mr. U. T—r, *who took the liberty to ask her the following Questions; viz. What age? What fortune? And what height she was?*

My height, Sir, in inches, is three times my years;
  My fortune their squares will both shew;
Put all these together, there then, Sir, appears
  The number exposed to your view.*   (*4494.
From which, Sir, determine the things you required;
  And then, if more favours you want,
As lovers of science I always admired,
  Those favors, perhaps, I may grant.

Answered by Mr. Tho. Baker.

Let x represent the lady's age, then her height (in inches) will be $3x$, and her fortune (in pounds) $= 10xx$, by the conditions of the question; from whence we have also given $10xx + 3x + x = 4494$; therefore $xx + 0.4x = 449.4$, and consequently $x = \sqrt{(449.44)} - .2\sqrt{(449.44)} - .2 = 21$. Hence the lady's age appears to be 21, her height five feet three inches, and her fortune 4410 pounds.

4.6 A Sampling of Mathematical Questions with Different Editorships

As we have seen, John Tipper, editor 1704–1713, was extremely careful that mathematical problems were set within the capacity and appreciation of his female readers. The following three problems from 1709 are typical examples:[19]

A gentleman has a garden of a rectangular form, the length 100 feet and breadth 80, and he wants to make a walk of equal width half way round to take up half the garden: What must be the width of the walk?

A Vintner has wines of 8, 5 and 4 groats per quart, and wants to make a mixture of 56 quarts, worth 22 pence a quart; how many different ways can this be done in whole numbers?

If thirteen tuns of claret cost nineteen pounds, how many pints can be had for a thousand crowns?

All of these problems can be solved within the realm of basic algebra and their content would appear gender neutral and non-offensive.

Henry Beighton, the following editor, 1714–1743, leaned more towards theoretical mathematical situations with more conditions imposed. He stressed scientific content involving experience appealing more to mathematicians and skilled technicians such as fellow surveyors. Problems become more masculine in their appeal:[20]

Required the sides and radii of the circumscribing circles of two regular polygons, the less of five sides and the other of seven, from the following data, *viz.* the less polygon is to the greater as 3 to 13, and the line subtending the angle formed by two adjacent sides of the pentagon is equal to the length of a pendulum which vibrates 61 times in a minute.

In gauging a spheroidical ale cask, I found the diameter of one head to measure 18·1 inches, that of the other 16, the bung diameter 20, and the distance between the two heads 20·6 inches, also, by the cask lying a little obliquely, I observed that the liquor just rose to, or touched the upper extremities of the two heads. Having noted these dimensions, I was informed that there were in the cask a ball of iron weighing 60*lb*. another ball of lead weighing 90*lb*. and a cube of box, a foot square. Pray what quantity of liquor was in the cask ?

A company of mathematicians, after emptying an elegant glass punch bowl, found themselves quite at leisure to contemplate its figure, &c. The inside had the form of an hyperbolic conoid, the transverse diameter of the generating hyperbola being 6·93 inches, and the conjugate 5·29 inches ; the form of the outside of the bowl was that of a cone generated by the revolution of the asymptote of the hyperbola, and the length of the outside was $8\frac{1.7.8.8.9.9}{4.8.0.8.8.8}$ inches ; also the depth of the bowl was 5·98 inches. Required the internal diameter at the brim, weight of the bowl, and what liquor it contained when full?

Problems now involving statics and dynamics began to appear, such as this one from the 1715 *Diary*:[21]

> Kind sir, I pray, can you to me declare
> A lofty tower's height within the air:
> I'll tell you how the height you well may know,
> Which in a problem unto you I'll show.
> If from the tower's height there should be laid
> A plain, whose surface fine and smooth is made,
> To meet the earth; three hundred foot and four
> From the foundation of this lofty tower;
> And then a body which in pounds doth weigh
> Just fifty six, you on the plain do lay;
> Just forty pounds will the same sustain,
> From sliding down on this descending plain.
> But, artist, I apply myself to you,
> (The tower's height) to calculate it true.

The plane geometry investigations became more complex, situations involving mechanics and solid and analytical geometry were introduced: mathematical disciplines and depth more readily available to men rather than women. The tone of mathematical problems became more occupational in nature and masculine related. However, the ladies still held their own in obtaining correct answers. Silvia, a consistent problem solver, supplied the correct answer to the 1715 problem, 310.27 ft., but in her 1716 reply she also gave the following retort as to the perceived disadvantage confronting her cohort female problem solvers:[22]

> Your towers lofty and sublime,
> Your problem rational and fine,
> Your methods just, I like the notion,
> Which join with numbers, weight and motion,
> But sure it is contriv'd to vex,
> Our uninstructed, softer sex;
> You try our weakness, search our flaws,
> By algebra and statick laws;
> Yet to untie your curious knot,
> Since 'tis a homely virgin's lot,
> Please to accept my kind, officious aid,
> Who am a rural and mechanick maid.

Robert Heath's contentious editorship lasted from 1745–1753. His problem situations became more fanciful and even more mathematically difficult.[23]

If the diameter of Sysiphus's cylindrical stone be two feet, which he continually rolls upon the surface of a semi-globular mountain, half a mile high : Quere what space will a spot on the convex surface of that stone travel through in rolling directly up and down the said mountain ? And what will be the time of its descent from the top by the force of gravity ?

Observing a horse tied to feed in a gentleman's park, with one end of a rope to his fore foot, and the other end to one of the circular iron rails, inclosing a pond, the circumference of which rails being 160 yards, equal to the length of the rope, what quantity of ground, at most, could the horse feed ?

A spider, at one corner of a semi-circular pane of glass, gave uniform and direct chase to a fly, moving uniformly along the curve before him : the fly was 30° from the spider at the first setting out, and was taken by him at the opposite corner. What is the ratio of both their uniform motions ?

To find the least number of guineas, which being divided by 6, 5, 4, 3, and 2, respectively, shall leave 5, 4, 3, 2, and 1, respectively remaining ?

These problems frequently require knowledge of the concept and use of fluxions and of higher analytic geometry. The last exercise on this list is an example of what would come to be known as the famous "Chinese Remainder Theorem" and was submitted by a lady. The Chinese mathematician Sunzi first recognized a systematic solution method to this problem in the third century; it was not formalized and published in Europe until the appearance of Gauss's *Disquisetiones Arithmeticae* in 1801. Here the solution is obtained through computation without the benefit of the Theorem. The "spider problem" is an example of a pursuit problem, amplified in importance by aerial combat during the First World War; such mathematical problems became a topic of later twentieth-century research.[24]

Thomas Simpson, editor 1754–1760, continued in a serious mathematical vein, offering problem challenges even to well-qualified mathematicians.[25]

If a straight, uniform, slender rod, or bar, of heavy metal, of a given length, be left to descend after being set leaning, in a given position, with its lower end (n) on the immoveable horizontal plane AB, and its upper end (m) full against the immoveable vertical plane AC (the lower end being at liberty to slide freely along the first-mentioned plane, while the upper end is descending), what will be the position of the rod when it shall cease to touch the said vertical plane ? how long will it then have been in motion ? and how far from the point A will the end (m) strike the horizontal plane ?

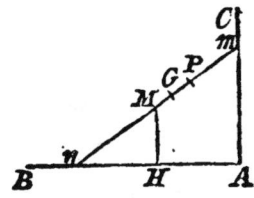

In a right-angled triangular field there are three trees, viz. one in each fence : The distance of the tree in the base from that in the hypothenuse, and from the acute angle adjacent, and of the tree in the perpendicular from the right angle, are all equal, and given ; and if lines be drawn from the tree in the base to the other two, those lines will form a right angle : The perpendicular of the proposed triangle is known to be the least of its kind (or that the data will admit of) : Hence you are desired to find the sides of the triangle, and to construct the same geometrically.

To assign the sum of the series $1 - \dfrac{1^2}{2^2} + \dfrac{1^2.3^2}{2^2.4^2} - \dfrac{1^2.3^2.5^2}{2^2.4^2.6^2} + \&c.$
by means of circular and elliptic arcs.

Simpson's problem-solving challenges exceeded Beighton's in difficulty. Problems in mechanics became more complex and the degree of mathematical analysis (calculus) required was quite high as exhibited in the last problem involving an alternating numerical series.

Edward Rollinson during his editorial tenure continued to maintain a high level of mathematical rigor in his mathematical problems series.[26]

The legs of a plane triangle being given equal to 30 and 40 respec_ tively ; and if a line be drawn from the vertical angle to the middle of the opposite side, the rectangle of the said line and base is a maximum ; required the bisecting line and base ?

Four spires standing directly north and south, and at the respective distances of 2, 3, and 4 miles from each other, were observed by a traveller, on a road tending to the north-east, the 1st and 2d, and also the 3d and 4th, appearing under equal angles, which they also did a 2d time, after travelling two miles farther on the same road ; required his distance from them at each observation ?

To determine the equation, the area, and the length of the curve, whose tangent and subtangent have always the same given difference (d).

To determine the nature of the curve, whose tangent terminated every-where by it and an indefinite right line CD, is a constant quantity $= 100$; also, to determine the length of the part thereof intercepted between its highest point and that point whose height, above the said right line CD, is $= 20$?

These problems have become distinctly academic, that is, textbook dictated. The situations are abstract and removed from practical reality.

Charles Hutton was the most experienced and respected mathematician to edit *The Ladies' Diary*, (1774–1818). He attracted young, trained mathematical talent to the challenges of his mathematical exercises and both developed and pressed their abilities with his problem-solving questions.[27]

To determine the position of the asymptote of the curve which is the locus of all the angular points formed by drawing three lines, from three given poles, so that the angle formed by the same corresponding two may always be bisected by the third.

To determine how far a man, who pushes with a force of 100lb. can introduce a sponge into a piece of ordnance whose diameter is five inches, and length 10 feet, when the barometer stands at 30 inches: the vent, or touch-hole, being stopped, and the sponge having no windage, that is fitting the bore quite close.

Required the dimensions of a cone, which if suspended by its vertex will vibrate as often in a minute, as it has inches in altitude?

If $1 + \frac{1}{2} + \frac{1}{6}$ be put $= A$, $\frac{1}{3} + \frac{1}{4} + \frac{1}{12} = B$, $\frac{1}{5} + \frac{1}{10} + \frac{1}{32} = C$, &c. then shall the arc of $90°$ be $= A - \frac{1}{4}B + \frac{1}{4^2}C -$ &c. Query the demonstration?

A quantity of matter being given, it is proposed to determine the figure of a solid of rotation made up of it, which shall have the greatest possible attraction on a point at its surface.

Military personnel from the Royal Military Academy were now submitting questions of a military nature and correspondents were asking for "Proofs". Note the "sponge and cannon" query. In this period of British history, the staff, officers and cadets at RMA would possess superior mathematical training as compared with common citizens even those of the gentry class. The mathematics considered has moved to a highly abstract plane.

Olinthus Gregory assumed the last editorship, 1819–1841. As a mathematician and RMA colleague of Hutton, he followed the same trend of promoting abstract problem solving.[28]

Sum the series $\operatorname{ch} a \operatorname{ch} \dfrac{a}{2} - \operatorname{ch} a \operatorname{ch} \dfrac{a}{2} \operatorname{ch} \dfrac{a}{2^2} \operatorname{ch} \dfrac{a}{2^3} + \operatorname{ch} a \operatorname{ch} \dfrac{a}{2} \operatorname{ch} \dfrac{a}{2^2}$

$\operatorname{ch} \dfrac{a}{2^3} \operatorname{ch} \dfrac{a}{2^4} \operatorname{ch} \dfrac{a}{2^5} -$ &c. to n terms.

If **any four points** be taken, then the following nine points lie in a conic **section.** The three intersections of lines passing through the points two and **two,** (1, 2, 3), and the six bisecting points, (4, 5, 6, 7, 8, 9).   For an hyperbola, **let each** point fall outside of the triangle formed by the other three ; for an **ellipse,** let *one* point fall inside the triangle of the other three ; for a parabola, **let one** point be at an infinite distance.

A shilling is thrown up and falls upon a chess-board : what chance is there of its lying upon one square, the side of each square being double the diameter of the shilling ?

Let there be given a point, a plane, and two spheres; to find a sphere which shall pass through the point, touch the plane, and also the two spheres.
This Problem, by a like method of reasoning, is immediately reduced to the viiith, where two points, a plane, and a sphere are given, and that by means of the vth Lemma. But if you choose to use the iiid Lemma, it will be reduced to the same Problem by a different method and a different construction.

These problems are highly academic and approaching the level of minor research endeavors. The "shilling drop problem" is a variant of the French gambling game, *"Le jeu de Franc-carriau"* where bets were made on the final resting position of a dropped coin upon a grid—did it touch a line or not? This is an example in geometric probability. The problem was analyzed by the French mathematician Georges-Louis Leclerc, Comte de Buffon in 1733. His results were published in 1777. Today, his work is recognized as the "Buffon Needle Drop Problem".

In scanning the above progression of problems and viewing their content and computational intent, it becomes clearly apparent that from approximately the period 1750 onwards the mathematics required to fruitfully engage in the mathematical problem competitions offered in *The Ladies' Diary*, one would require advanced training in mathematics and a high level of personal skill. The problems had seriously moved beyond the talented self-trained or home-schooled amateur to attempt and even for much-determined and mathematically competent ladies. From entertaining diversions, the *Diary*'s mathematical problem posing evolved into extremely challenging exercises.

5

Mathematics, Education, and Women in Eighteenth- and Nineteenth-Century England

What were women's opportunities to study and know mathematics?

In examining the issues surrounding the appearance of *The Ladies' Diary* as a venue for mathematical problem solving and expression, especially one initially instituted and shaped by women, several questions arise: mainly, "Why now?" and "How?" In a large sense, the second of these questions has been answered in our considerations of the almanac as a literary and scientific genre, market conditions, and the background and personalities of its various editors and contributors. But as a periodical, and a very popular one, devoted to problem solving in the unraveling of word puzzles, enigmas, and the finding of solutions for mathematical exercises, one must ask "Why was the general public drawn to these features and their required tasks?" Once again part of this query, the one concerning enigmas and their intellectual and societal attractiveness, has been satisfied; the appeal of the mathematical challenges remains to be more fully explained. Just where did they fit in?

5.1 Emerging Mathematical Priorities

No movements, intellectual or social, arise spontaneously; they must be stimulated and shaped. So it was with the societal interest and enthusiasm for mathematical problem solving that the appearance of *The Ladies' Diary* served the British public. The historical period in which the *Diary* flourished in the British Isles, 1704–1841, has often been described as "mathematically depressed" and a period of mathematical and scientific indecision. This malaise has been primarily attributed to the break with continental science and mathematics due to the Newton-Leibniz controversy over the "discovery" of the calculus and due priority. Another theory for the situation has been attributed to the lassitude and lack of action on the part of the Royal Society—they did not actively promote mathematics and science. Despite some truth in both claims, there were significant mathematical activities taking place during this time span. The problem with discerning and discussing "mathematical progress or developments" depends on how one "sees" and evaluates mathematics. Certainly from the time of the Enlightenment through the middle of the nineteenth century, English mathematical involvement in such areas as: surveying, cartography, navigation, astronomy, instrument making, fortification and military logistics, timekeeping, banking and insurance reckoning, greatly increased. To best understand the perception of mathematics in modern English society, that is, post sixteenth century, one must look at the rise and impact of the initial, egalitarian-directed, stimuli for an appreciation of number and its power.

By the time of the reign of Queen Elizabeth I (1558–1603), Britain had become an island nation with an emerging Empire of scattered colonies, possessions, and dependents. Its maritime supplied and maintained the oceanic lifelines: insuring national security, providing territorial expansionism, and preserving imperial dominance. But, "Were the ships' officers adequately trained in the newly resurrected sciences of astronomy and geometry in order to compete with the other advancing European nations?" Profit, political prestige, and military dominance were at stake. England had to obtain and maintain navigational superiority of the seas. Elizabeth was requested to supply financial support for studies in astronomy and geometry at Cambridge and Oxford. These subjects were the basis for navigation.[1] She refused, perhaps quite wisely, believing that financial support of matters pertaining to navigation and ultimately commercial expansion should be supplied by the private sector. To a large extent, this strategy had already been in place and was working.

The mathematician Robert Recorde (ca. 1510–1558) had been employed in the 1540's to research and teach navigation to the members of the Muscovy Company as had the Queen's own scientific adviser, John Dee from 1550–1583. Thomas Digges (d. 1597) served his patron Robert Dudley, Earl of Leicester in advising on navigational matters as well as defense. The young and promising mathematician Thomas Harriot (ca. 1560–1621) was hired by Sir Walter Raleigh, ever the commercial adventurer, to aid him with matters involving navigation and cartography.

A wealthy merchant and politician, Sir Thomas Smith, later to become the first Governor of the East India Company, sponsored a series of free public lectures on mathematics and navigation. The Company's resident mathematicians, Thomas Hood and Edward Wright delivered these lectures. The work of such mathematicians employed by commercial companies would soon be available to wider and more general audiences through further series of lectures and their publication of popular textbooks.

Among these books, now published in the English vernacular, was Robert Recorde's *The Grounde of Artes, Teachings the Worke and Practise of Arithmetic*, (1543) [**Rec43**].[2] Recorde combined his writing with a strong conviction of the popular utility for mathematics, prefacing his book with an imaginary dialogue between a learned master and his doubtful, or inquisitive, "scholer" who seeks a justification for learning mathematics. These first ten pages were later singled out as "The Declaration of the Profit of Arithmetic" and communicated to attract adherents to the subject.[3]

The "numbering" referred to and advocated in *The Grounde of Artes* is that system employing the "new," innovative, "user friendly" Hindu-Arabic numerals and their computational algorithms. Robert Recorde who wrote for "the young and rude" attempted to impart an appreciation for mathematics with a warning: "yf nombre be lackynge, it maketh men dumme, so that to most questions, they must answer mum." The book became very popular, going through a total of forty-seven separate printings.

Now several books by the father and son team, Leonard Digges (1520–1559), a surveyor, and Thomas Digges, an astronomer, also advocated the use and applications of mathematics in daily life. *Tectonicon* (1556) [**Dig56**] expounded on geometry, surveying and the use of measuring instruments; *A Geometrical Practise Named Pantometrie* (1571) promoted the same theme; and *An Arithmetical Militare Treatise named Stratioticos* (1572) further demonstrated the uses of mathematics in warfare, military engineering, and architecture.[4] Also appearing at this time in English history was the first English language translation of *Euclid's Elements, The elements of geometrie of the most ancient philosopher Euclides of Megara* (1570). The translation was completed by Henry Billingsley (d. 1606), a prosperous London businessman and civil servant.[5] Highlighting this historic edition was a fifty-page "Mathematicall Praeface" by John Dee (1527–1608?), a prominent and mysterious figure of the time, mathematician, astrologer, and occult philosopher and, most importantly, scientific advisor to Queen Elizabeth I.

In contributing to Billingsley's work, Dee lauded the power and versatility of "Numbryng", noting a Divine intention:

> All thinges do appeare to be Formed by the reason of Numbers. For this was the principall example or patterne in the minde of the Creator.[6]

Dee's "Preface" to *Euclid* was directed mainly at literate common folk, urging the study and use of mathematics. He included examples of mathematical principles that his readers could easily identify with and perform. This declaration was particularly influential among the rising class of artisans and craftsmen, the "mecanicians," of his time.

Within this onslaught of the popularizing literature on number and mathematics, other notable texts appeared, *Cocker's Arithmetic: Being a Plain and Familiar Method Suitable for a Full Understand of that Incomparable Art* [**Coc44**], Edward Cocker (1677)—the "Incomparable Art" being arithmetic. Although the book first appeared after Cocker's death, it became the most popular arithmetic book of its time, going through 130 editions and served as a standard English school text for the next century. Peter Ramus's *The Way to Geometry* was translated from the French into English and published in 1636. The extended title read "... Being necessary and useful for, Astronomers, Geographers, Land-meaters [Surveyors], Sea-men, Engieners, Architecks, Carpenters, Paynters [Painters], Carvers, etc."

The writings of Recorde, Diggs, Cocker, and Ramus helped remove the veil of mystery that had surrounded numerical computation and geometric renderings. At last almost anyone could read numbers and do arithmetic. Benjamin Wardhaugh, in his bibliometric analysis of the context of English books during the period 1473–1800, notes authors' trends of using the broader term "mathematics" increasingly in preference to such specific words as: "arithmetic", "geometry" and "astronomy."[7] By the time of the eighteenth century more than a quarter of the books examined spoke of "mathematics." This practice of recognizing the encompassing field *mathematics* as a distinct intellectual focus and endeavor is particularly noteworthy, marking a psychological and sociological transition.

This growing popular interest in mathematics spawned a variety of supporting activities. Public lectures on mathematics were given in such places as coffeehouses and bookshops. In 1696, Charles Cox, a London Brewer, established a series of free, public mathematics lectures. At first, these presentations concerning merchant accounting and navigation were given in the busy London district of Southwark; later they were moved to the Marine Coffee House, Birchlin Lane, near the Royal Exchange. In the period 1698–1704, the recognized scientist and well-known Newtonian, John Harris, delivered a series of lectures at this coffee house. Other coffee houses and pubs adopted this practice. Around England, local organizations were formed to support mathematics and science learning: the Spitalfields Mathematical Society, 1717;[8] the Spalding Gentleman's Society,

1717; the Manchester Mathematical Society, 1718; Northampton Mathematical Society, 1721; and the Newcastle Literary and Philosophical Society, 1793. Joseph Middleton, a teacher of mathematics on ships and a surveyor, founded the Spital-fields Society in order to teach mathematics and navigation to sailors; however, its members became a mixture from the local community. These were working men's clubs, where participants would gather once a week to enjoy a pipe and a pint and discuss mathematics and solve mathematical problems from sources such as *The Ladies' Diary*. Eventually, some such societies established public lecture series. Ladies were permitted and attended these science and mathematics lectures. Women also quietly formed their own system of female-oriented study groups.

Both the realized approachability and the usefulness of numbers and mathematics had a profound impact among the common people. Number now meant power, profit, and success. Quantification and measurement were viewed in a new light, as both controlled acquisition and ownership. Accurate merchants accounts provided records allowing for comparisons, revealed deficiencies, and isolated excesses, improving management and dictating the allotment of resources. Precise measurements bounded land, defined ownership, navigated ships, insured schedules, constructed buildings, affirmed weights and capacities, and promoted a state of well-being. Recorded numerical data allowed for deliberation and the rendering of decisions. Now with an enhanced understanding of their world, common people began to "do" and "use" mathematics. These "Practioners" surveyed the land, directed the ships, kept the merchant's accounts, gauged the barrels and kegs, taught mathematics, made instruments, drew maps, mixed the medicines, amalgamated metals, and performed dialing tasks.[9] Furthermore, they shared their knowledge through the writing of textbooks, almanacs and "ready reckoners," mathematical tables complied for easy consultation. An example determining the interest on a loan was John Seller's *Pocket Book of Tables*, 1677. See Figure 5.1. Seller even numerically referenced commodities such as wine and bread. See Figure 5.2.

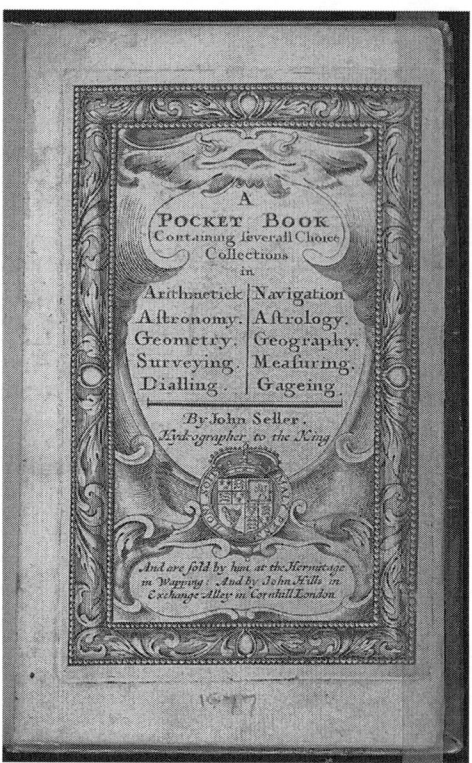

Figure 5.1. As the general British populace became more numerate conscious; various mathematics tables were published for consulting purposes. One such book was John Seller's *Pocket Book of Tables*, 1677. It contained a series of comprehensive numerical reference tables.

The rising knowledge and appreciation of mathematics demanded more supporting resources in the manner of formal, systemized, instruction.

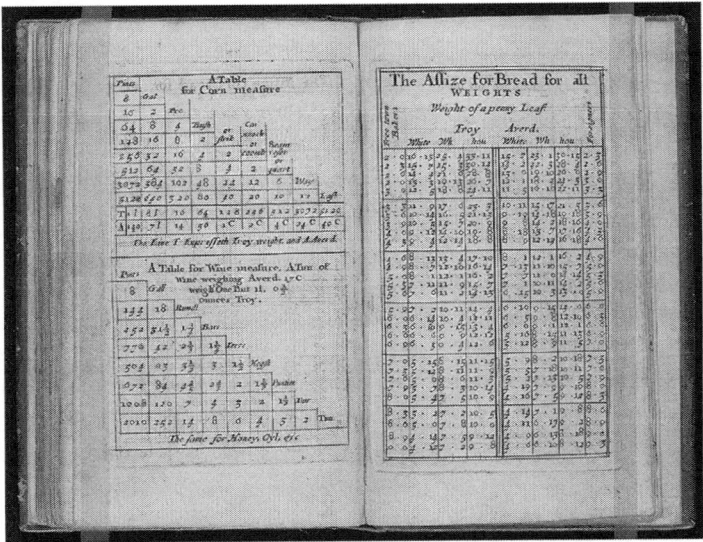

Figure 5.2. Even bakers were guided by numbers dictating the composition and size of a loaf of bread. Source: Beinecke Rare Book and Manuscript Library, Yale University.

5.2 A New Educational Thrust

To accommodate this new awareness, individual teachers, writing and arithmetic masters, and reckoning masters set up practice. One could seek out a master and, for a price, become literate and/or learn the science of land survey or ocean navigation. Such tuition could be subscribed to on an individual lecture basis or the student could choose to live with the mathematics master for a prolonged period of time to better receive adequate training. Some such situations would result in the founding of proper academic academies. These "progressive" schools taught the new sciences and competed with existing traditional institutions that still functioned with a classical curriculum promoting Latin and Greek. Education was still sharply divided along social class and gender. There were few and intellectually limited schools for girls.

Sir Thomas Gresham, a merchant, financier and founder of the Royal Exchange, funded the establishment of Gresham College in 1598. His College was intended mainly to train craftsmen, mariners, and members of the working classes in the use of the applied mathematical sciences, especially in the areas of astronomy and navigation. Its meeting rooms served as an active forum for professors, *philomaths*, mathematical practitioners, mariners, and navigators to exchange knowledge and experiences.

During her rule, Queen Elizabeth's scientific advisor, John Dee, had urged the adoption of maritime studies focusing on the principles of mathematical navigation. His pleas were to no avail. Samuel Pepys (1633–1703), noted English diarist, had spent much of his life in administrative service to the Royal Navy, rising from the position of a member on the Naval Board to that of Chief Secretary of the Admiralty, an honor he was to retain through the reign of King Charles II and into that of his successor, King James II. Pepys lobbied for better funding and improvements of the Royal Navy. He knew of a Dutch military engineering school and French, Spanish, and Portuguese schools of navigation and felt that England needed such a facility for the training of its seamen. However, he was not the only one harboring this concern; two wealthy and influential merchants, Robert Clayton and Patience Ward, envisioned beginning such a school attached to the existing Christ's Hospital in London. When presented with their idea, Charles II approved and issued a charter in 1673 proclaiming the founding of the Royal Mathematical School, RMS, at Christ's Hospital.[10] See Figure 5.3. The designation of the term "Mathematical" in the school's title signifies an official recognition of the discipline's increasing importance. The mathematical practitioner and respected surveyor Jonas Moore (1617–1679) was appointed governor of the school. There was some controversy as to the appropriateness and rigor of the mathematics curriculum. Details on the specific curriculum are vague: arithmetic was to be taught as far as "the Rule of Three"; algebra up to the ability to obtain solutions for quadratic equations; Euclid's *Elements* were to be studied as well as Trigonometry, "Plain & Spericall." Students would be taught the use of maps, globes, and various measuring instruments. They would have to be able to determine longitude at sea. Moore, an experienced mathematician, began to compile a text specifically for the RMS students; however, he died before finishing the work. The work, *A New Systeme of the Mathematicks*, was completed by others, credited to Moore, and published posthumously in 1681 [**Moo81**]. Ironically, the cost of the text was too expensive for it to be adopted at the Christ's Hospital school.

Figure 5.3. A contemporary announcement memorializing the founding of the Royal Mathematical School, 1673. Credit: Christ's Hospital, London: the Mathematical School. Etching in two colours by B. Green, 1793, after himself, 1775. Credit: Wellcome Collection. Attribution 4.0 International (CC BY 4.0)

The RMS, Christ's Hospital, was a success and soon similar mathematical schools were founded: at Greenwich in 1684 and at Rochester in 1708. Encouraged by the movement, several private, Huguenot-run academies were established where instruction was also focused on mathematics and navigation. The Huguenots were a group of French Protestant immigrants who fled Eighteenth century religious persecution in their country. Skilled craftsmen and artisans, they had a high respect for science and education and, upon coming to England, many became personal tutors and eventually founded their own schools. Private enterprise also saw the flourishing of other "mathematical schools." Their names proudly advertised their purpose: Sir Joseph Williamson's Free Mathematical School at Rochester (1701) and Neal's Mathematical School at Fleet Street, in London. All of these institutions taught the mathematics of navigation.[11]

While the scientific modernizing of the Royal Navy received a special priority, it soon became obvious that the other mainstay of British national security—the army—also required scientific-based reforms. In 1741 the Royal Military Academy, RMA, Woolwich was formally established outside London. This facility was intended for the purpose of:

Instruction of the raw and inexperienced people belonging to the
Military branch of this office in the several parts of Mathematics
necessary to qualify them for the service of the Artillery, and the
business of the Engineers.[12]

The standardization and improvement of artillery during the latter part of the
eighteenth and into the nineteenth centuries saw the development of a science of
ballistics and the use of instruments for the distant-sighting of cannon. Counter-
point to these advances were the improvements in the design and construction
of fortifications enabling them to better withstand the effects of prolonged bom-
bardment. Highly qualified mathematics masters were sought out as instructors.
From 1754 onwards, the editorships of *The Ladies' Diary* were occupied by RMA
mathematics masters. These teachers at RMA, themselves contributors to *The
Ladies' Diary*, formed a critical mass of influence and direction for the development and reform of nineteenth-century British mathematics.[13] Most of them
were from humble origins, self-educated, and devoted to teaching. Their educational efforts were from below and upward, not theorizing on mathematics but
rather exposing and promoting mathematics as an understandable and empow-
ering tool for the common people.

The remaining military academy with a link to *The Ladies' Diary*, and which
also exerted a recognizable mathematical influence on the educational scene, was
the Royal Military College at Sandhurst, instituted in 1799. Thomas Leybourn
joined the staff as a mathematical master. He had been a frequent contributor to
The Ladies' Diary and held Charles Hutton in great respect. Both of these men
eventually served as editors of the *Diary*. William Wallace (1768–1843) joined
Leybourn in the teaching of mathematics at Sandhurst in 1803, and James Ivory
(1705–1842) arrived the following year. Both were highly accomplished Scottish
mathematicians.

As for the existing universities, Oxford and Cambridge, mathematical reforms would come more slowly. It must be remembered that we are dealing
at this time with a very stratified society separated by class and gender. It was
a time of societal transition where an onrushing modernism was confronting a
classical traditionalism. Societal norms to be followed, especially those involving education differed greatly. The channels for official education were limited to
the upper classes or those men destined for service to the Church of England. For
these few, schooling may have begun at home with a tutor or at a boarding school
that offered a classical education with language training in Latin and Greek. If
the student sought a specific education, he (remember education was limited to
males) would seek out an appropriate professional such as a lawyer or medical
doctor to study under. Otherwise the student could enter the military or go on to
university, Oxford or Cambridge, where classical education would continue. In

1570, some of the colleges at Cambridge adopted reforms in the teaching of the mathematical sciences. St. John's and Granville and Caius colleges began to offer courses in arithmetic, geometry, perspective, and cosmology but these offerings were not met with much student enthusiasm. To help promote reform and attract highly qualified professors, Henry Savile (1549–1622) in 1619, founded two honorary chairs, one in astronomy, the other in geometry. Despite such efforts, Oxford clung to its classical roots, resisting any scientific modernization.

At Cambridge, the situation was very similar; mathematics was viewed with suspicion and disdain. While in 1663, Sir Henry Lucas, Member of Parliament, founded a Chair of Mathematics, the Lucasian Chair, at the University, it was also strongly felt that the study of mathematics was inappropriate for a "Gentleman." John Wallis (1616–1703), England's most eminent mathematician before Isaac Newton, lamented the poor mathematical training he had received as an undergraduate at Cambridge. In 1635 upon writing about his experiences, he noted that he had no one to direct him and that:

> Mathematicks were scarce looked upon as Academical Studies but rather Mechanical; as the business of Traders, Merchants, Seamen, Carpenters, Surveyors of Land, or the like; and perhaps some Almanack-makers in London.... And among more than two hundred students in our college, I do not know of any two which had more Mathematicks than I, which then but very little...[14]

Finally he directed anyone desiring to learn mathematics to go, not to the university but rather, to London. He was saying that if one desired to learn mathematics, he should go to those people who were using it, the Practioners.[15]

John Locke (1632–1704), England's most respected Enlightenment philosopher, found fault with the universities' focus on humanistic scholasticism. He denounced the use of Latin rather than the English language for instruction and favored more teaching of modern, relevant subjects such as science and mathematics. In his *Some Thoughts on Education*, published in 1693 [**Loc93**], Locke was not concerned with the common people; his advice was directed at the gentry. Concerning mathematics, he counseled:

> Merchant's Accompts, though a Science not likely to help a Gentleman to get an estate, yet possibly there is not any thing of more use and efficacy, to make him preserve the Estate he has...I would therefore advise all gentlemen to learn perfectly Merchant's Accompts, and not to think it a skill, that belongs not to them, because it has received its Name, and has been chiefly practiced by Men of the Traffick.[16]

The merchant may be despised but his ability to do arithmetic and keep accounts should be valued and emulated!

By the beginning of the nineteenth century, the curriculum emphasis at Oxford and Cambridge began to diverge. Both institutions still functioned to serve

"Gentlemen" but both the natures of the gentlemen and their specific educational needs changed. Oxford retained its classical stress on rhetoric and logic, while Cambridge, the academic home of Isaac Newton, adopted more studies in mathematics and physics. These scientific disciplines were justified as they "taught young men to reason." But the teaching of classical geometry remained sacrosanct even under challenges as to its mathematical rationale. In 1813, a group of mathematically talented undergraduate students at Cambridge formed the Analytical Society, an organization devoted to promoting a stronger analytical approach to the teaching of English mathematics.[17] In 1819 the Cambridge Philosophical Society was founded at the University. This organization provided a further stimulus to the improvement of science teaching and research at the institution.

The setting of formal examinations also influenced conditions for the learning and teaching of mathematics at the universities. The written examinations given to distinguish and honor graduates in mathematics at Cambridge were known as the Mathematical Tripos, the name being derived from the three-legged stool on which an examination candidate sat. Initially devised by the University's Senate House in 1740, Tripos were set for the testing of several disciplines but the Mathematical Tripos was the first to be instituted at the University and became the most prominent. A high performance/score on this examination earned the honorary title of "Wrangler," while successful, but lesser achievements warranted titles of "Senior and Junior Optimes." Becoming a Wrangler on the Mathematical Tripos garnered high respect. Even though some Wranglers rationalized the Tripos as a test of general academic excellence designed "to make good members of society—not to extend science and make profound mathematicians," it did have that effect. The new academic emphasis was not appreciated by all undergraduates at the University, as noted by an anonymous "B.A." writing in a student journal in 1756. He complains:

> Mathematics is the standard to which all merit is referred: and all other excellences, without these are overlooked and neglected: the solid learning of Greece and Rome is a trifling acquisition: and much more so, every polite accomplishment: in short, if you will not get all Euclid and his diagrams by heart, and pore over Saunderson 'till you are blind as he was himself, they will say of you... 'tis all over with you! You are ruined!...[18]

Still from the middle of the nineteenth century onwards, many of England's most eminent mathematicians have been Wranglers.

This new emphasis at Cambridge, of judging and rewarding mathematical excellence through the use of a written problem-solving examination, the Tripos, was at this time, so radical an action that it deserves a closer examination. While the official justification for such an innovation is based on its demonstration of "reasoning ability" in the candidate another possible rationale might lie in the existing national trend of mathematical problem solving. Is it a mere coincidence that the format and content of a mathematical Tripos closely approximates the sets of mathematical exercises offered in *The Ladies' Diary*?

This series of educational reforms—the establishment of mathematics schools, military academies, and changes in university curriculum—while extensive, were male orientated, initiated, and directed. Few concerns for the desires and needs of women were considered.

5.3 A Woman's Exposure to Mathematics

While John Tipper intended the *Ladies Diary* to serve a wide female audience from shop girls, to household servants, to young ladies of leisure, practical circumstances intervened to limit this audience. Perhaps some ladies' maids and governesses, much to their credit, did participate and benefit in the exercises of the *Ladies Diary*, but the majority of respondents were young ladies of leisure, for unhindered periods of time provide the best/necessary conditions for mathematical deliberations and problem solving. They were the daughters of the newly rising class of prosperous merchants, industrialists, clergymen, professionals, and landed gentry, the "Ladies" of the new era. While the time-honored aristocratic class strove to preserve their order, this new genteel class strengthened their positions of power and influence by appropriate and convenient intermarriage. In a sense a marriageable daughter was a commodity, one to be carefully groomed for her aspiring position as the wife of a Gentleman. Ideally, she was to be genteel in her demeanor; pleasant in conversation but not too opinionated; bearing the necessary social graces to entertain at gatherings of her peers, to amuse, and entertain without being intrusive: to recite poetry, sing or play an instrument; to be accomplished in some ornamental arts such as water painting or embroidery; maintain the household, with the required servants; and most importantly, to bear children. An female author's comments on an English woman's life in the seventeenth-century is equally relevant to the following century:

> Women were not borne to read authors, and censure the learned, to compare lives and judge of virtues, to give rules of morality, and sacrifice to the Muses. We are willing to acknowledge all time borrowed from family duties misspent... If sometimes it happens by accident that one of a thousand aspires a little higher, her fate commonly exposes her to wonder, but adds little to esteeme.[19]

Most of this training, education in subservient domesticity, was given at home by parents, or for the more wealthy, a governess. The principles of a Christian morality were strongly embedded with the practice of Bible study.

Of course most British women were aware of the economic and intellectual changes of their world and the world "at large." The century-long efforts to advance an appreciation and use of mathematics were not lost on women. They read the editorials, attended the lectures, and acquired some of the new mathematical texts. They too wished to participate in these new opportunities and challenges. Instruction and education were required. The services of writing mistresses who also taught arithmetic could be procured. Unfortunately, but telling, is the fact that little information is available on the existence and functioning of these teachers. Obviously, they were an educated group of ladies available to teach other females seeking knowledge and skills usually denied their sex. Pages from an existing copybook, 1688, belonging to fifteen-year-old Mary Serjant demonstrate the learning expectations. See Figure 5.4. Mary is receiving instruction in writing, calligraphy and arithmetic by a Mrs. Elizabeth Bean. Note the decorative and ornamental setting in which the arithmetic is presented hardly promoting its utility.[20]

Figure 5.4. Pages from Mary Serjant copybook, 1688, demonstrate her competence in penmanship and arithmetic. Penmanship or calligraphy as practiced with a quill pen at this time included artistic flourishes. These extravagances dignified and advertised the refinement and social standing of the practitioner. Source Beinecke Rare Book and Manuscript Library, Yale University

Fortunately another copybook from one of Mrs. Bean's students is also available for examination. The student is a Sarah Cole who, in 1685, recorded a lesson, most probably dictated, which goes beyond a mere writing exercise to instill a value for arithmetic and computation:

Arithmetic is the Art of Computation

By Numbers which bring many Consolation

Those who true reckoning from false Discern

Arithmetick Let Them Completely Learn

By the Merchant and Man of Trade

By Ignorance or Skill are marr'd or made

Yet in this Art Therse none that's so accurate

As all Its Excellencies To Compute[21]

Arithmetic's (spelled two ways above) association with trade and commerce was firmly embedded in the seventeenth-century psyche.

A sympathetic and knowledgeable governess or a progressive relative might also tutor young ladies in mathematics. Women could, and did, attend public lectures on the new sciences and their projected applications. The wives of merchants and scientific craftsman would become acquainted with their husband's increasing concern with numbers, measurement, and computation.

The questions concerning a proper education of an English woman were the subject of opinions and pronouncements. Bathsua Makin who had served as a tutor to the daughter of Charles I urged an expansion of educational opportunities for women. In her *An Essay to Revive the Ancient Education of Gentlewomen*, 1673 [**Mak73**], she counseled:

> ...women ought to be brought up to a comely and decent carriage, to their needle, to neatness, to understand all those things that do particularly belong to their sex. But when these things are competently cared for, and where there are endowment of nature and leisure, the higher things ought to be endeavored after.[22]

As for "carriage," a young lady entering a boarding school soon found herself confined within corsets and braces to adjust her posture to that appropriate of a lady. This physical constriction provides a suitable metaphor for the intellectual confinement to which she was also subjected. The "higher things" referred to were such disciplines as languages, painting, dancing, singing, and the keeping of accounts—assets of genteel domesticity.

In the middle of the eighteenth century, Elizabeth Montague (1718–1800), socialite and patron of the arts, was instrumental in establishing the "Blue Stocking Society." This society, consisting of both men and women, would meet periodically and discuss literature and the arts. Its purpose was to expose women to a broader intellectual world, using conversation as a means to acquire knowledge. It can be said that as a result of such interactions, more British women became involved in writing and literature; however, the effects on their scientific broadening were minimal. The term *"Blue Stocking"* remains as a reference, usually derogatory, to female intellectuals of this period. Despite a rising chorus of women's voices advising on their intellectual and social status, it would be a male concern for reform that would warrant more apt attention. These recommendations for the modernization of female education would come from Erasmus Darwin (1731–1802), a respected physician and natural philosopher, later to obtain fame as the grandfather of Charles Darwin. In 1797, he published *A Plan for the conduct of female education in boarding schools* [**Dar97**], in which he proposed a broadening and standardization of the curriculum. Yes, he believed that mathematics and sciences should be taught but limited to latter years of a young girl's schooling.[23]

There were some academies for young ladies that followed Darwin's suggestions and offered studies in mathematics. Two such examples were a school run by a Mrs. Florian, at Leylonstone, Epping Forest and Bryan House, Blackheath founded by Margaret Bryan who was recognized as a learned scholar in the sciences. See Figure 5.5. In 1797 she issued copies of her *Lectures on Astronomy and Mathematics*, notes from her classroom presentations. Astronomy was one of the few sciences approved for young ladies, perhaps due to its celestial associations. One copy found its way into Charles Hutton's hands at Woolwich. So impressed was he with her work that he sent her a note of admiration on January, 6, 1797 expressing his feelings:

> I have read over your lectures with great pleasure and the more so,
>
> To find that even the learned and more difficult sciences are thus
>
> Beginning to be successfully cultivated by the extraordinary and
>
> elegant talents of the female writers of the present day.[24]

This endorsement helped to promote her school and enrollments noticeably increased.

Figure 5.5. A 1797 engraving by the English engraver William Nutter (c. 1759–1802) depicts the female educator and social reformer Margaret Bryan with her daughters. Source: Wikimedia Commons.

As for Mrs. Florian's academy, some specifics on the mathematics taught is described: "The principles of arithmetic are now demonstrated, and its use in housekeeping concerns and in the occurrences of life... The elements of Geometry and Trigonometry are also taught as far as requisite for a perfect intelligence of the principles of Astronomy, of the geographical knowledge of our globe and of Natural Philosophy..." Her rationale justifying the teaching of mathematics reflects on its still tenuous acceptance as a subject for the consideration of young ladies.

As for the status of young ladies' academies in general, Jane Austen, in her novel *Emma*, 1815, describes female boarding schools as institutions:

"... where a reasonable quantity of accomplishments were sold at a reasonable price, and where girls might be sent to be out of the way, and scramble themselves into a little education, without any danger of coming back prodigies."[25]

5.4 Contemporary Testimony on the Subject of "Women and Mathematics"

Women studied, discussed, and participated in the rising effort to problem-solve and use mathematics as demonstrated in *The Ladies' Diary*. Mindful of their men's efforts to join together in "study-group" clubs and associations to learn and practice mathematics, they also banded together in "us too efforts" in the learning of mathematics.

Correspondent/contributors to the *Leeds Correspondent* in the 1821 and 1822 issues provide some insights and opinions of ladies' mathematical activities at this time. From 1815 to 1823, James Nicolas of London published this annual periodical described as "*A Literary, Mathematical and Philosophical Miscellany.*" In volume III, 1821, of this periodical under "Miscellaneous" contributions and following an entry entitled "A Dissertation on Forks" is found an item, "Female Mathematicians" submitted by a "Dedascalus," a schoolmaster from the village of Mytholme ["**D21**]. He proclaims how the ladies of his region exhibit a "relish for mathematics and scientific instruments" and are particularly "enamoured by the study of mathematics." This movement of "mathematics mania" began when a learned lady from London settled in the area. Knowledgeable in geometry and the sciences, she began to teach her new lady friends geometry, "these charming mysteries." They took to this subject readily and soon replaced their gatherings of tea and cards with ones focused on mathematics problem solving. They adopted "philosophical and mathematical amusements at their select parties." At first tentative in their mathematical abilities, their intellectual soirees were limited only to ladies but as their confidences increased they also invited gentlemen to the gatherings. Recognized as a "lover of the Mathematics," "Dedascalus" was invited to attend one of these gatherings. He joined an assembled group of thirteen ladies and four other gentlemen. Their meeting began with tea and an hour of agreeable discourse before the assembly divided into separate problem-solving groups of four or five individuals seated at separate tables. Each individual had a slate on which to compute and every table was presided over by a "President" who would inspect and approve the mathematical work undertaken. The members contributed a shared library of resource and reference books. Among the selection present were: *The Ladies' and Gentlemen's Diary* and Leybourn's *Repository*. While at the session they served as a source of problems, at another time, one might find *The Ladies' Diary* serving the same function. For a period of two hours, individuals and groups commenced upon the solution of chosen mathematical problems. Some efforts were cooperative, and the whole encounter enlivened with amicable and helpful conversations. As "Dedascalus" noted:

> ...even those who cannot produce a correct answer derive a considerable benefit, and add to their stories of knowledge, by observing the labours

of others. Their emulation is likewise excited and their taste improved by beholding the various ways in which proper solutions are produced, and by choosing out of them those which are the most concise and elegant.[26]

As a schoolmaster he commented on the limited education provided for women and their obvious latent abilities but took great pride and was "delighted with their employment," in mathematical activities.

A rebuttal also entitled "Female Mathematicians" appeared in the next yearly edition, written by a Liverpool bookseller, "Bipliopola" who did not approve of women doing mathematics. As he related, while he was entertaining a group of friends, the *Leeds Correspondent* bearing "Dedascalus's" declaration arrived by post from London. Several of the male company began reading the contribution describing the contemporary female fashion of mathematical problem solving. As the narrator declared, its contents caused them much consternation while their female companions listened with delight and approval. Within less than a year of this episode, many of the ladies present became "tolerantly expert mathematicians and newly fledged philosophers." "Bipliopola" exclaimed "O unhappy me!" as his wife was among this group. He was sorry she gave up the playing of cards for the pleasures of mathematics. Justifying his distress further, he explained:

> But nothing can be more prodigiously anomalous and unnatural than for the wife of a tradesman to be puzzling her brains with the solution of a difficult problem when the dinner should be smoking hot on the parlour table, and her family partaking of it.... To the most obtuse understanding therefore, it must be evident that the study of Mathematics is entirely incompatible with the occupations of one on whom the daily cares of and minute concerns of a family have developed.[27]

While admitting that the *Correspondent*'s article stimulated sales of his mathematical books, he lamented and condemned the notion of women doing mathematics, ending his contribution with the question:

> Why then should the ladies be suffered to encroach on our own studies [men's], or to intermeddle with masculine concerns?[28]

This exchange of opinions and experiences illustrates the cultural/social and intellectual conflicts still surrounding the status of British women and their association with mathematics at this period of history. Even towards the end of the century, in 1884, the prejudice against women doing mathematics in Great Britain was obvious as conveyed by the mocking lyrics of Gilbert and Sullivan's comic opera *The Princess-Princess Ida*:

In mathematics, Woman leads the way:

The narrow-minded pedant still believes

That two and two make four! Why, we can prove,
We women, household drudges as we are—
That two and two make five or three or seven;
Or five-and-twenty, if he case demands![29]

6

The Ladies' Diary as a Facet in the Mathematical and Scientific Transition of the Era

What intellectual value did the *Diary* possess?

6.1 Fulfilling a Mathematical Need

Now, with a better appreciation of the popular mathematical movements that began in sixteenth-century England with the appearance of Recorde's and the Digges' arithmetic and geometry books and the continuing momentum for the learning and utilization of mathematics that ensued in the following century, we can more fully understand the impact and influence of *The Ladies' Diary* at the time it appeared on the scene. *The Ladies' Diary* appeared at a critical juncture of societal mathematical involvement. It encouraged, taught, and provided "feedback" in mathematical problem solving to a receptive audience. It was the first major, popular periodical to have such a feature. The *Diary* satisfied a pressing need and its success both instigated and strengthened the development of an English mathematical problem-solving genre. It served as a ready model for a variety of competing problem-solving periodicals that appeared in its wake. Following the *Diary*'s example, even newspapers began to feature "Problem Columns" with the same format and rationale. One such column began in 1828 in the *York*

Courant by Thomas Tate, a highly respected mathematics author and educator. An editorial accompanied the initiation of this column noting:

> That no paper in this city, had a corner devoted to Mathematical Recreations, we very readily opened our columns to their communications.[1]

At the beginning of 1830, another editorial appeared responding to a challenge as to the usefulness of mathematics. It strongly affirmed the newspaper's support of mathematics, stating in part:

> But give us leave to ask what has been the means of raising our nation to the pitch of opulence and civilization to which it has at present arrived, in the richness and beauty of its manufactured goods; the symmetry of its edifices, its bridges, its canals; the immensity of its wind, water and steam engines, etc.? The answer to this question may be "All the sciences have contributed", which is very true; but let us add that the science of Mathematics, aided by Chemistry, has done more than all the rest put together...[2]

Tate's mathematical column ran weekly until September 1846.

While initially publishing mathematical exercises as an entertaining feature, *The Ladies' Diary* introduced and groomed its audience to the appeal and utility of mathematical problem solving, the power of doing mathematics. These revelations would be duplicated through additional hosts of popular literature.

6.2 Acquiring New Mathematical Skills and Understanding

Both the *Diary*'s mathematical questions and the enigmas were puzzles that demanded solutions. Such solutions were obtained through the application of directed mental efforts, through a similar process of problem solving. This feature, call it either "the unraveling of a puzzle" or "problem solving," is what made the *Diary* a popular and long-lived periodical. It guided and taught the reader to:

(1) Recognize the problem.

(2) Gather the clues or hints pertaining to an answer.

(3) Employ the correct technique, logical or mathematical, to fit the clues together.

(4) Obtain an answer.

(5) Test and judge the "correctness" of the answer.

(6) Apply the answer.

In understanding how to work with mathematics, the third through the fifth steps in this process are most important. A problem-solver needs feedback: "Is the answer correct?" "Will it work?" One of *The Ladies' Diary*'s strongest assets was to do exactly this by supplying complete, worked-out solutions for all its mathematical problems. In some cases, it presented alternate solution procedures and references, perhaps exposing a reader and correspondent to new, broader and useful information. The solutions and explanations extended knowledge. They taught!

This question was posed in 1715, Silvia's response in verse was previously examined in Chapter 4, p. 42, below the answer is given in its computational form that appeared in 1716. Unlike the problems posted by John Tipper, under the editorship of Henry Beighton problems in mechanics began to appear, a true challenge to ladies who were not supposed to have an acquaintance with such a science. An additional solution was added by Hutton and published later in Leybourn's collection.[3]

No. 12, 13.]   MATHEMATICAL QUESTIONS.   4i

VI.   QUESTION 46, *by Mr.* Ed. Elphick.

Suppose a smooth inclined plane from the top of a lofty tower meets the level ground at the distance of 30½ feet from the bottom of the tower, and suppose a weight of 40lbs. will sustain a weight of 56lbs. when placed upon the inclined plane : What is the height of the tower ?

Answered by Silvia.

By a known principle in mechanics, the accelerating velocity, or weight of bodies on an inclined plane, is to their accelerating velocities or weight in their perpendicular descent, as the sine of the angle of inclination, to the radius ; or as the perpendicular to the length of the plane, considered as an hypotenuse; and therefore in this the proportion of the perpendicular, and the length of the plane, being given as 40 to 56, or as 1 to 1·4. Let the perpendicular sought be $= x$, the hypotenuse will be 1·4x, and the base 30½ feet $= a$, then $1·96x^2 - x^2 = 0·96x^2 = a^2$ by 47 Euclid 1. and by division and evolution,

$$x = \left(\frac{aa}{0·96}\right)^{\frac{1}{2}} = 310·27 \text{ feet required.}$$

Additional Solution.

It is evident that this problem may be easily *Constructed* thus : Make a right-angled triangle whose hypotenuse and perpendicular are 56 and 40, or 7 and 5, or any two numbers proportional to these ; then make another similar triangle whose base may be 30½; and its perpendicular will be the height of the castle required.

And from the same principle, the *Calculation* may be given without algebra, thus : Since $\sqrt{(7^2 - 5^2)} = \sqrt{24} = $ the base of the first triangle to which the other is similar, we shall have as $\sqrt{24} : 5 ::$

$$304 : \frac{304 \times 5}{\sqrt{24}} = \frac{304}{\sqrt{96}} = \frac{760}{\sqrt{6}} = 310·2687 = \text{the castle's height.   H.}$$

4

This question was published in the 1733 *Diary* under Henry Beighton's purview. Two solutions are supplied but one must wonder at the nature of a "Grubean Lady." This female geometer prefaced her solution with a verse greeting her "geometric friend," Sam Ashby. The author of the problem, Samuel Ashby was so favorably impressed with this lady's response that in a letter he challenged her to use her "leisure" time to solve his new problem, a "Prize Problem" 1735.

ᴵᴵ. QUESTION 170, *by* Mr. Sam. Ashby.

If upon each leg AB and BC, including the right angle, be drawn a square BD and BE; and the lines DC and EA, which cut the said legs at F and G. I say, BF and BG are equal, and are each a mean proportional between the segments AF and CG; that is, as AF : FB :: FB : CG, &c. Querc, the demonstration geometrically?

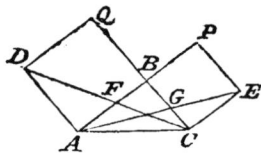

Answered by Mr. Rob. Fearnside.

It is plain, by similar triangles, that CB + AB : AB :: BC : FB. Again CB + AB : BC :: AB : GB. Permutando CB + AB : AB :: BC : GB. Ergo FB = GB.

For the other part of the demonstration; by similar triangles AB (= AD) : AF :: BC : BF; and AB : BG :: BC (= CE) : CG. Therefore, *ex equo* AF : BG :: BF : GC. *Q. E. D.*

A Grubean Lady's answer to the same.

The triangles APE and ABG, being similar, it is AP : PE :: AB : BG, (by Euclid 4. VI.); that is AB + BC : BC :: AB : BG, because all the sides of a square are equal. In like manner the triangles CQD and CBF, being similar, it is QC : CB :: DQ : FB; *i. e.* AB + BC : BC :: AB : FB. Therefore FB is = BG. Again, the triangles DAF and FBC, and the triangles ABG and ECG are similar; therefore (DA or) AB : BC :: AF : FB; and AB : BC (= CE) :: BG : GC. Whence AF : FB :: BG (or BF) : GC (by Euc. 11. V.) *Q. E. D.*

5

x. QUESTION 797, *by Mr.* Alex. Rowe, *of Reginnis.*

Required the nature and quadrature of the curve into which a flexible wire 11·3052 feet long must be bent, so that a ring of heavy metal being put thereon, and the wire revolved about an axis at right angles to the horizon, with a given velocity, the said ring shall rest equilibrio ; the abscissa being to the ordinate as 3 to 2.

Answered by Mr. John Whitton.

Let AM be the required curve, AP its axis, PM and BD ⊥ AP, and DC ⊥ to the curve ADM at D.   Then, since the angular velocity is given or constant, the centrifugal force, in direction BD, at any point D of the curve, will be as BD, which resolves into the two forces BC, CD ; of which the latter is destroyed by the curve, being perp. to it ; therefore the former, or subnormal BC, is a constant quantity, being directly opposite and equal to the force of gravity ; and consequently the curve will be a parabola.   And, its length being given, by means of the general expression for its length, we obtain AP = 9, PM = 6; and hence the area APM = 54.

G 2

The challenge problem, untaken by the lady.

If in any plain triangle, as ABC, you draw lines from each angle through any point E (o) within the triangle, till they cut the opposite sides, as CA, AB, and BC.   The rectangles of the alternate segments at those sides will be equal.

viz. Ab × ca × BC = bc × aB × aA.

The demonstration of this curious proposition is required ?   6

This problem was posed in 1782 with Charles Hutton as editor. Note Rev. Hellin's extensive commentary, which also continued with two Scholia, the whole comprising a brief but informative dissertation on the problem:

The same by the Rev. Mr. John Hellins.

Let ADC be the curve into which the wire must be bent, D any point in it, and DE ⊥ AB the axis. Call AE, x ; DE, y ; AD, z ; $32\frac{1}{6}$, f ; 3·1416, p ; and the time of one revo- 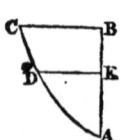 lution, in seconds, n. Then, by mechanics, $f\dot{x}z^{-1} =$ the force of gravity in direction\ of the curve at D ; also $4p^2yn^{-2} =$ the centrifugal force at D in direction ED, and therefore in direction of the curve it will be $4p^2y\dot{y}n^{-2}z^{-1}$; and this must be equal to the above opposite force arising from gravity, therefore $n^2f\dot{x} = 4p^2y\dot{y}$, and the fluents give $n^2f = 2p^2y^2$, or $ax=yy$, (putting $fn^2 \div 2p^2 = a$), which is the equation of a parabola. Now the length of the curve of a parabola, whose abcissa is 3, and semi-ordinate 2 feet, is $= \sqrt{10} + \frac{1}{4}$h. l. $(3 + \sqrt{10}) = 3·7684$ feet ; hence $11·3052 \div 3·7684 = 3$; therefore $3 \times 3 = 9 = $ AB, and $3 \times 2 = 6 = $ BC. Consequently $fn^2 \div 2p^2 = a = 36 \div 9 = 4$, and $n = p\sqrt{(2a \div f)} = 1·5667$ seconds $= 1\frac{17}{30}$ seconds very nearly.

Corollary 1. The centrifugal force at c, the end of the bounding ordinate, or top of the curve, $4ppy \div nn = 2fy \div a,$ is $= 3f,$ three times the force of gravity.

Corollary 2. The force by which the ring is urged in direction of the curve at c, both by gravity and the centrifugal force, is $f \div \sqrt{(1 + a \div 4x)} = 3f \div \sqrt{10}.$

Corollary 3. The pressure of the ring against the curve at c, is $= \sqrt{(ff + 9ff)} = f\sqrt{10},$ or the pressure is to the weight of the ring as $\sqrt{10}$ to 1.

7

John Hellins was a schoolmaster and popular author of textbooks.

The following question was asked in 1826; note a citing of Euler's work in the answer and the referencing of scholarly journals: *Mémoires de l'Académie Impériale des Sciences de St.Pétersbourg,* vol 9 (1803+) and Ferrusac [sic] *Bulletin des Sci. Math.* The second of these journals is actually *Bulletin des Sciences Mathématique, Physiques et Chimiques,* published by the French aristocrat Baron d'Andebard Ferussac during the period 1824 to 1831. The reference most probably applies to the first volume, 1824, revealing that the *Diary*'s editor at this time, Olinthus Geogory had access to the latest scientific journals from the Continent. How many other of *Dia*'s readers would have such an opportunity is questionable.

IV. QUEST. (1452); *by Investigator.*

Find *three* numbers, such that their sum shall be a square, and the sum of their squares a biquadratic. Then find *four* numbers, and afterwards *five* numbers, possessing the same properties.

IV. QUESTION ; *ans. by Mr. W. Almond, Birtley ; and Mr. Mason, Scoulton.*

First, let $\sqrt{4x-5}$, $2x-1$, and x^2-2, be the three req. numbers ; the sum of their squares being x^4 a biquadratic. It remains that their sum be made a square, viz. $x^2+2x-3+\sqrt{4x-5}=a\square$: this it will be if $-3+\sqrt{4x-5}=1$, or $4x-5=16$, or $x=\frac{21}{4}$. Therefore, the numbers are $\frac{64}{16}$, $\frac{152}{16}$, $\frac{409}{16}$; or in integers 64, 152, and 409.

Secondly.—Let $\sqrt{6x-6}, x-2, x-1$, and x^2-1, be the four numbers, the sum of their squares being x^4, a biquadratic ; their sum $x^4+2x-4+\sqrt{6x-6}$, will be a square, if $-4+\sqrt{6x-6}=1$, or, $x=\frac{31}{6}$. The Numbers, then, are $\frac{180}{36}$, $\frac{114}{36}$, $\frac{152}{36}$, $\frac{925}{36}$, or 180, 114, 150, and 925, in integers.

Lastly.—Let $\sqrt{4x-12}, x+1, x-1, 2x-1$, and x^2-3, be the 5 req. nos. the sum of their squares being again x^4. Their sum, $x^2+4x-4+\sqrt{4x-12}$, will be a square, if $-4+\sqrt{4x-12}=4$, or $x=19$. Hence the Numbers are 8, 20, 18, 37, and 358.

Cor. By a method exactly similar may any number of numbers be found possessing the same properties.

°.° Many of the solutions, and especially those of *Messrs. Maffett, Riddle, Rutherford,* and *Thompson,* were very neat and satisfactory; but none exhibited the simple elegance of the above.

†₊† In vol. 9, *Mem. Acad. Imp. Sci. Pet.* an interesting posthumous paper of *Euler's* is given, on this subject. He finds, also, the *minimum* numbers in the several cases; sometimes, however, obtaining two numbers alike. Thus, for 3 numbers he presents, for the *min.* solution, 49, 64, 8; for 4 numbers, 137, 88, 32, 32; for 5 numbers, 89, 72, 32, 16, 16. See also *Ferrusac, Bulletin des Sci. Math.* iii. 276.

8

The published solutions, detailed in themselves, supplied the readers learning opportunities. Many authors referenced contemporary texts and other relevant published works. More examples of problems and their feedback solutions can be found in Appendix C at the end of this book. Thus in this manner, these published solutions and their commentaries supplied a network of mathematical information. *The Ladies' Diary* greatly contributed to, and even accelerated, the mathematical movements taking place in Great Britain.

6.3 The *Diary* as a Provider of Scientific Facts

The British people at the turn of the eighteenth century, while complacent within their sense Imperial dominance, were also consumed with a growing curiosity about the forces changing their lives. New scientific, economic, and social movements were providing information that displaced traditional beliefs, especially those concerning the place and potential of an individual within the larger society. The advancing middle class sought out information to satisfy their self-education efforts. A variety of printed materials—broadsheets, newspapers, pamphlets, and periodicals—appeared to solve this need. Almanacs including *The*

Ladies' Diary, were very much a part of this trend. Books, *per se*, were still too expensive for the average consumer. Supplying relevant and attractive information to an eager audience became a major goal of publishers, especially the Company of Stationers. Under John Tipper's initial design, the *Diary* became a vehicle and forum of scientific information. Tipper's "Geography Paradoxes" prompted the reader to explore the world in an attempt to find places cryptically described:

> There is a certain place on the Globe, of considerable Southern Latitude That hath the greatest and least Degree of Longitude.

> There are Three remarkable places on the Globe that differ both in Longitude and Latitude and yet all lie under the one and same Meridian.

> There is a certain Island in the Baltic Sea to whose inhabitants the Body of the Sun is clearly visible in the morning before he riseth and likewise in the Evening, after he is set.[9]

Simple diagrams, although they were accompanied by explanations, provided visual information and understanding derived from their configurations and labeling. For example, in the 1706 *Diary*, the universe is discussed and the differences between Ptolemaic and the Heliocentric systems of planetary rotation are clearly illustrated.[10] See Figures 6.1(a) and 6.1(b).

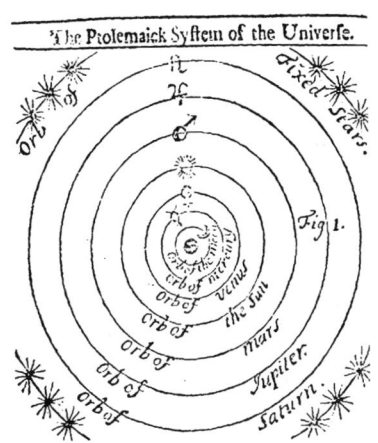
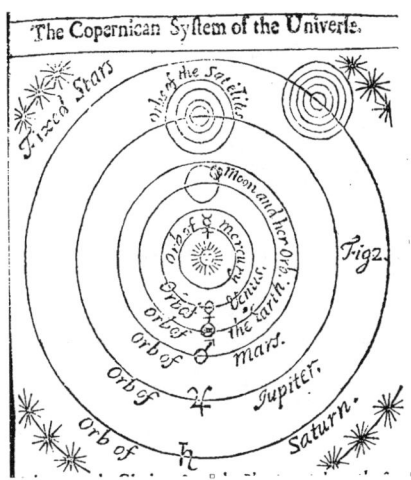

Figure 6.1. Although the ancient Greeks had theorized that the universe was solar centered, that is that the earth and other planets rotated about the sun and later, during the European Renaissance, Johann Kepler, Copernicus, and Galileo established this theory as a scientific fact. Many people, due to cultural or religious beliefs, did not except this heliocentric theory. For such people, the earth-centered or Ptolemaic concept of heavenly behavior, still seemed true. Thus, in the era of scientific enlightenment sweeping Great Britain in the seventeenth and eighteenth centuries, the question of planetary behavior deserved attention. In the 1706 *Diary*, the universe is discussed and the differences between the Ptolemaic, Figure 6.1(a), and the Heliocentric, Figure 6.1(b), systems of planetary rotation are clearly illustrated. Source: Harry Ransom Center, University of Texas, Austin.

Similarly, in 1709, a "Mystery of the Heavens" is clarified in a sketch explaining the motion of comets. Its annotation denotes a then recent scientific discovery by the astronomer Edmund Halley.[11] See Figure 6.2.

But perhaps the best, prolonged feature of *The Ladies' Diary* that focused on information was the question and answer column begun by Robert Heath. In the 1749 issue, Heath introduced "The Ladies Oracle or Querest," allowing individuals to pose open questions for responses. By the 1754 edition of the *Diary*, he had embellished the title of this column to "Questions for Exercise and Improvement of the Invention, Reason and Judgement." This column became quite popular in the freedom and scope of questions and answers it solicited. During the life of the *Diary*, the questioning emphasis moved from mostly that of "lovelorn maidens," concerns on love and marriage, to broader queries on science and the world at large. Some typical questions were:

1772—How many different Passions of the mind do tears proceed?

- All agree that every Event is directed by an over-ruling Power, for what Reason are many things attributed to chance?

- What is Chance?

1786—How high must the sun be to render a rainbow invisible?

1807—Whence comes the dew we see in the morning?

Does it fall from the air above, or rise out of the earth?

1825—What is the best method of completely freeing iron or steel?

From magnetism?[12]

Poetic exposition was still very popular in all submissions to *The Ladies' Diary*: In 1826 a Mr. James Herdson, using such a literary form, asked:

Dear Ladies, I've this question put

In hopes of your solution-

Is tea injurious, or not,

To an English constitution?[13]

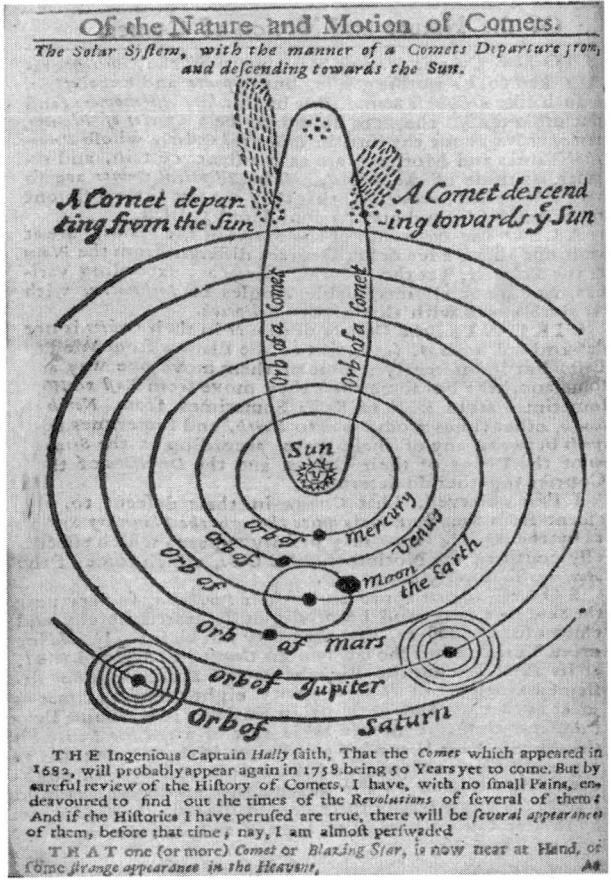

Figure 6.2. In 1705, the English astronomer Edmond Halley published *Synopsis Astronomia Cometicae*. In this book, he used Isaac Newton's gravitational theories to chart the paths of two dozen comets. Within his research, Halley speculated that three, assumed separate, comets observed in 1531, 1607, and 1682 were actually the same object and he predicted that it would reappear again sometime in late 1758 or early 1759. It did return in 1758, marking a great victory for British science. Source: Harry Ransom Center, University of Texas, Austin.

An Answer provided by a Mr. Joseph Wilkinson:

The China herb will never harm

An English constitution;

If, of good cream and butter'd toast

There be no destitution.

Mr. J. E. Barrat advises: I think that the infusion from genuine

black tea, provided it is not too strong, it is not injurious to the constitution.

But the chemical process for converting it into *green*

tea, renders it highly injurious to the animal system, as is also

every kind of adulterated tea, in which usually, there is a great

portion of sngar [sugar?] of lead.

One is left wondering just how this information/advice affected our inquisitive tea-drinker. Furthermore, "What would our last advisor think of modern flavored teas?" By the way, he is correct in his caution of processed green teas at that time, as they contained high levels of lead.

6.4 Instructional Essays and Dialogues

The publication of instructional essays in *The Ladies' Diary* was dependent on and varied with the interests, motivation, and priorities of its editor. John Tipper's dislike for astrology led him to offer a series of essays scientifically explaining the night skies from his 1706 discussion of the "Nature of Eclipses of the Sun," to "On the Nature and Motion of Comets," 1709, and on to the "Constellations of the Zodiac." In this last contribution, aided by illustrations, Tipper appears as the consummate teacher, leading his pupil, the reader, to a recognition and understanding of each Zodiac constellation. Consider how he explained the constellation *Taurus*. See Figure 6.3 and the following explanation:

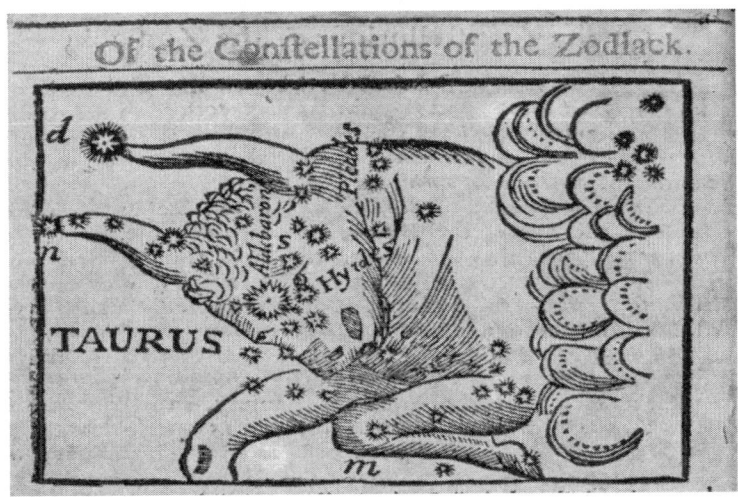

Figure 6.3. Taurus the bull rages through the heavens. Source: Harry Ransom Center, University of Texas, Austin

THE *next Constellation of the Zodiack is Taurus*, the
BULL, denoted by the Ancients with a *horned Head of a
Bull*, thus 8. This *Image* consists of 48 Stars, whereof
one is of the called his *South Eye*; it is also called by
the *Arabians, Aldebaran*. Another in the *Tip of
his North Horn* at (*d*) is of the *Second Magnitude*, that at
South Horn, (*m*) the Star in the bend of his Knee, & *c*.

He continues to explain the whole configuration in such a manner. Then the
schoolteacher/editor instructs his readers in accomplishing their own future celestial observations:

By what *means or ways* must I *know these Stars in the Heavens?*
I *answer.* There are divers ways or helps to know them;
One of which I shall now mention, and reserve the *rest* till
by and by.
1.THE *First way then to know the Stars is, By having the Picture or Representation of the Constellations, by you, and knowing one or more Stars
in the Heavens of the same Constellation, by comparing Stars in the picture or Figure, and those in the Heavens, and considering their Situation,
Distances and Magnitudes in the one, you may easily find out those in the
other, ...* [14]

The following editor, Henry Beighton, was quite occupied in a variety of activities, and by now, the *Diary* was receiving an excess of acceptable correspondence, more than could be published in the limited space available. He chose not to publish extended essays and articles in favor of the readers' contributions. His successor, Robert Heath actively promoted the new theory of fluxions and in two series, 1746–1747 and 1751–1752, he contributed articles on the theory and concept of fluxions.[15] Following Heath, Thomas Simpson took the reins as editor. While Simpson was an active author, he did not see fit to contribute or publish any lengthy articles, again relying on the correspondence of readers to fill the *Diary*'s pages. The next two editors, Rollinson and Hutton, followed the same policy. But under the editorship of Olinthus Gregory, the situation changed; the repeal of the Stamp Act in 1835 reduced the costs for the *Diary* and allowed for the addition of more pages. Beginning with the 1835 issue, he added an "Appendix" devoted to articles on mathematics and science. In total through the remaining editions of the *Diary*, Gregory included twenty-one such articles. He explained his intentions for this innovation in that "...the supplementary sheets, thus given, if carefully preserved, in a very few years will form a rich repository of instruction for mathematical students." See Figure 6.4. The majority of articles were on geometry and the behavior of mathematical series. Beginning in 1811 and onward, the very versatile and prolific William Horner (1786–1837) made many contributions to the *Diary*. Horner's "A New Method of Solving Numerical Equations of all Orders by Continuous Approximation," a paper presented to the Royal Society in July of 1819, appeared as an Appendix article in the 1838 edition.[16] The 1837 collection of articles included one on the "Investigation of a Recoil Engine," a concession to the industrial movement.[17] The works of foreign authors were also included or referenced: the continentals, Giacomo Riccati, P. Fermat, J. Strum, and L. Necker and an American from Long Island, New York, Charles Gill. Hence, by the middle of the nineteenth century, the *Diary* began to take on aspects of a professional journal. Gregory's enriching feature would be continued in *Dia*'s successor, *The Lady's and Gentleman's Diary*, under the heading "Mathematical Papers."

Figure 6.4. The cover of the last issue of *The Ladies' Diary* which featured a new Queen and offered "Valuable Mathematical research." Source: Haiti Trust, Public Domain

While some features of *The Ladies' Diary* changed or diminished in reader participation, throughout its lifetime as an independent journal it remained a source of information. Beyond the continued verbosity of its entertaining enigmas, the *Diary* carried on John Tipper's initial goal of providing its readers with scientific facts and explanations.

7

"*Dia*" as a Mathematical Testament

Did the *Diary* reflect and support the mathematical reforms taking place during the span of its publication?

7.1 The Winds of Change

The period during which *The Ladies' Diary* existed as an independent periodical, 1704–1841, was a time of mathematical ferment and gestation in the British Isles. Concerted educational efforts, both formal but mostly informal, increased the popular appeal and appreciation of the utility of mathematics. The establishment of Mechanics Institutes for working-class people in the beginning of the nineteenth century and the founding of the University of London in 1826 added further momentum to this mathematical awakening. More and more people knew some mathematics and used it. The immediate acceptance and adoption of the *Diary* as a mathematical problem-posing and solving vehicle reflects this movement.

As for ferment, several specific mathematical concerns also surfaced or became relevant during this same period: the clarification of Newton's theory of fluxions, a reorientation towards the analytic processes of mathematics rather that those that stressed a synthetic approach, the seeking of a more precise method of determining longitude at sea, the intellectual relevance as to the content and manner in which Euclidean Geometry had been traditionally taught, and the need to reestablish a better knowledge and working relationship with continental developments in mathematics and science. Newton's calculus, method of fluxions, was originally published in Latin with many concepts remaining obscure in their theory and explanations.[1] The foundational basis of Newton's theories were open to question, a situation soon to attract the attention of Bishop George Berkeley (1685–1753) who became a vocal and articulate critic of the new mathematics.[2]

In 1736, John Colson published his translation and annotation of Newton's theories in *The Method of Fluxions and Infinite Series* [**New36**]. Other mathematicians also published texts attempting to clarify and mathematically strengthen the theory of fluxions. Most notable in this effort were Brook Taylor, Colin Maclaurin, and Thomas Simpson. From 1736–1758, twelve treatises on fluxions were published.

In addition to these theoretical studies, there appeared several editions of Newton, specifically composed for women. In 1742 an English translation of the Venetian polymath, Francesco Algarotti's *Il newtonianesimo per le dame* [Newtonianism for the Ladies], 1737, appeared as *Isaac Newton's Theory of Light and Colours and his Principles of Attraction made familiar to ladies in several entertainments.* Elizabeth Carter (1717–1806), the eminent female intellectual and noted "Blue Stocking," completed the first English translation of Algarotti's work [**ACN57**]. Carter's exposure to Newtonianism for British women was very popular and went through several editions. Later in 1801, Colson's translation of Maria Agnesi's *Instituzeoni Analetiche* [Analytical Institutions] [**Col44**], 1748, was published. In the Preface to his translation, which he entitled "The Plan of The Lady's System of Analyticks," Colson clarifies his intentions for the female reader:[3]

> attempt it's tranflation, though I well knew how unequal I was to the tafk. I confefs I alfo entertained fome diftant hopes, that it might excite the curiofity of fome of our *Englifb* Ladies; that it might raife an emulation in them, a laudable ambition to promote the glory of their country, with a generous refolution not to be outdone by any foreign ladies whatever. They want no genius or capacity for the fciences, and have undoubtedly as good abilities as the Ladies of *Italy*. They feem only to want to be properly introdueed into thefe ftudies, to be convinced of their ufefulnefs and agreeablenefs, and to

In 1770, the British Navy's Board of Longitude offered a prize of 20,000 pounds for a means or method for obtaining an accurate measure of longitude at sea. Existing approximation methods had proven unsatisfactory and resulted in shipwrecks and the deaths of seamen. Many mathematicians and scientists became involved in this lucrative quest. Newton had unsuccessfully tried his hand at the problem. Other mathematicians such as Humphrey Ditton and William Whiston succumbed to the challenge, as did Robert Hooke, Edmund Halley, and even the architect Christopher Wren. Ditton and Whiston's futile efforts: *New*

Method for discovering the Longitude both at Sea and Land, 1714, were met with the ridicule of the satirist and social commentator Jonathan Swift in his ribald lyrics to "Ode to musick on the longitude."[4] Robert Heath offered a scheme which was dismissed by the Royal Astronomer Nevil Maskelyne who himself was an unsuccessful seeker after the prize.[5] As a result of this rejection, Heath added Maskelyne's name to his long list of perceived enemies. Even Charles Hutton was involved in preparing mathematical tables for the Navy's navigational needs. But the competition was not limited to men, or Englishmen alone. Two women, Elizabeth Johnson and Jane Squire, submitted proposals and the Swiss mathematician, Leonhard Euler received a partial prize of 300 pounds for his assistance with the compilation of lunar tables. At the time, an anonymous "Ballad of Gresham College" called attention to the situation:

The Colledge will the whole world measure,

Which most impossible conclude,

And Navigation make a pleasure

By fynding out the Longitude.

Every Tarpaulin [Tar] shall then with ease

Sayle any ship to the Antipodes.[6]

The main prize was eventually won by a mathematical practitioner, a Yorkshire carpenter and clock maker, John Harrison, who made a durable and accurate marine chronometer, designated as H-4. Now sea-travel time, traversed from a fixed reference line set through Greenwich, England, could be reliably employed to find the longitude of a position at sea.

Geometry instruction particularly on the first two books of Euclid had been a mainstay of British classical education for centuries. Isaac Newton relied on geometry in formulating his *Principia.* The subject was almost sacrosanct as a vital part in the training of an English gentleman. But it was taught more as a mental discipline—train the mind; learn logical reasoning—than as a mathematical tool. In the beginning of the nineteenth century with the discovery of non-Euclidean geometries and the growing popular interest in the utility of mathematics, a new academic emphasis on an analytic approach to the understanding of mathematics called into question the rationale for teaching geometry.[7] Aware of continental geometry teaching reforms, British authors began to produce more "user-friendly," pedagogically appealing, practical geometry texts. Two examples in this new movement were Robert Wallace's *A Treatise on Geometry*, 1831, and Thomas Tate's *Principles of Geometry*, 1848. The second half of the nineteenth century saw an estimated seventy-three different books on geometry used for instruction in England.

Although the transition of English mathematics towards a more analytical approach is often attributed to the efforts of the Cambridge Analytical Society,

British mathematicians such as Woodhouse, Toplis, Ivory, Wallace, Playfair, and Hutton were well aware of continental mathematical developments, took advantage of them, and urged their colleagues to follow suit. Writing in 1789, John Playfair chided his friend, the Royal Astronomer Nevil Maskelyne and other fellow mathematicians for their mathematical isolation:

> He is much attached to the study of geometry, and I am not sure that he is very deeply versed in the late discoveries of the foreign algebraists. Indeed, this seems to be somewhat the case with all the English mathematicians; they despise their brethren on the Continent, and think that every thing great in science must be forever confined to the country that produced Sir Isaac Newton.[8]

In the 1805 issue of the *Philosophical Magazine*, John Toplis (1774–1857), a respected mathematics teacher and translator, made the following similar appeal to the British mathematical community:

> It is remarkable, that amongst the very few men who still pursue mathematical studies in this country, a considerable part, instead of being dazzled and delighted by the wonderful and matchless powers of modern analysis, still obstinately attach themselves to geometry. It is a science, perhaps, of all others, from the clearness and accuracy of its proofs, the most proper to be taught young men, that from the study of it their reasoning faculties may be improved; but at the same time, as a science, it is confined in its application, feeble, tedious, and almost impracticable in its powers of discovery in natural philosophy. But what is called analysis possesses boundless and almost supernatural powers in its application to science; and they pretend to despise it, and obstinately to grovel amongst a few properties of surfaces and solid bodies, part of which were discovered by means of analysis, denotes a very narrow and prejudiced mind.[9]

In 1808, John Playfair lamented the prevailing stagnant status of British calculus as compared to continental advancements:

> The calculus of sines was not known in England till within a few years. Of the method of partial differences, no mention, we believe is yet to be found in any English author, much less the application of it to any investigation. The general methods of integrating differential or fluxionary equations were all to them unknown; and it could be hardly said, that the more difficult parts of the doctrine of Fluxions, any improvement that has been made beyond those of the inventor. At this moment when we now write, the treatises of Maclaurin and Simpson, are the best we have in the fluxionary calculus, though such vast improvements have been made by the foreign mathematicians, since the time of their first

publication. These are facts, which it is impossible to disguise; and they are of such an extent, that a man may be perfectly acquainted with everything in mathematical learning that has been written in this country, and yet find himself stopped at the first page of the works of Euler and D'Alembert.[10]

In his efforts to promote the work of the French analysts, John Topis published an English language translation of the first two books of Laplace's *Celestial Mechanics* in 1814. James Ivory (1765–1842) had already written on volume three of this work in 1809. The same year William Wallace interpreted Legendre's work on elliptic integrals for a British audience. Although he often disapproved of the techniques used by his mathematical peers across the channel, Robert Woodhouse (1773–1827) admired the work of Lagrange and strongly advocated the use of differentials in his *The Principles of Analytical Computation*, 1803. So while the efforts of the Cambridge Analytical Society did draw attention to the perceived deficiencies of English mathematics, the situation was long recognized and being acted upon.[11] When many of the undergraduates involved in the Analytical Society latter became faculty at Cambridge University, they led the way and once again Great Britain excelled in mathematical accomplishments.

The decades spanned by the existence of *The Ladies' Diary* experienced changing moods as to the directions in which mathematics should move. Did the periodical's contents reflect these societal opinions in any way? Did the emphasis on problems posed and solved during this period change accordingly? Were the controversies or movements evident in textual contributions—editorials, letters to the editor, or published essays?

7.2 *The Ladies' Diary*: Trends and Influences

Several researchers have surveyed the mathematical exercises considered in the *Diary* and categorized them according to mathematical significance and societal relevance such as trade, business, work experience, and household needs, and the life-style of the times. Some broad but penetrating comments on the influence of the 0 (the concept of zero as a numerical limit) were made in 1808 by John Playfair (1748–1819). Playfair was a Scottish mathematician, recognized as a mathematical reformer and a highly respected educator. These comments were part of a book review that he published for the local journal, the *Edinburg Review*. Playfair's contribution was actually focused on Laplace's five-volume *Mécanique Céleste* [**Lap23**] but in it he also speaks to the existing state of mathematics in England as he then saw it, singling out *The Ladies' Diary* for recognition:

> A certain degree of mathematical science, and indeed no inconsiderable degree, is perhaps more widely defused in England, than in any other country in the world. *The Ladies' Diary*, with several other periodicals

and popular publications of the same kind are the best proofs of this assertion. In these, many curious problems, not of the highest order indeed, but still having a considerable degree of difficulty, and far beyond the mere elements of science, are often to be met with; and the great number of ingenious men who take a share of proposing, and answering these questions, whom one has never heard of anywhere else, is not a little surprising. Nothing of the same kind, we believe, is to be found in any other country... The geometrical part... has always been conducted in a superior style; the problems proposed have tended to awaken curiosity, and the solutions to convey instruction, in a much better manner than is always to be found in more splendid publications.[12]

While Playfair's comments affirm the mathematical influence of *The Ladies' Diary*, his omission of the mention of women problem solvers is particularly noteworthy. His blinders of social convention and the role played by women in shaping the *Diary* still matched existing perceptions. When John Playfair died, and his library was dispersed, a collection of *The Ladies' Diaries* were found among his literary possessions.

Here, our discussion will focus on *The Ladies' Diary* as a vehicle illustrating and communicating the movements discussed above. Was it a passive witness or an active agent in the developments?

Thomas Leybourn (ca. 1769–1840), a mathematics master at the Royal Military College at Sandhurst and a professional associate of Hutton, published *The Mathematical Questions Proposed in the Lady's Diary and Their Original Answers* in 1817. This work in four volumes covered the *Diary*'s problem content from 1704 to the year 1816. The problems cited below are referenced by the listing numbers employed in Leybourn's 1817 edition of *The Mathematical questions Proposed in The Ladies' Diary, 1704–1816*; beyond this date range, the numbering will still follow Leybourn's system, if it were extended; otherwise it follows the numbering of the specific host issue of that *Diary*. See Figure 7.1. Since over time, several numbering systems have been employed in the *Diaries* and their compilations, this ordering would seem reasonable. Readers who wish to know more about the solutions offered for the cited problems are referred to Leybourn's *Mathematical Questions* which is available online from the Hathi Trust Archive or, if retrievable, to the relevant issue of *The Ladies' Diary*.

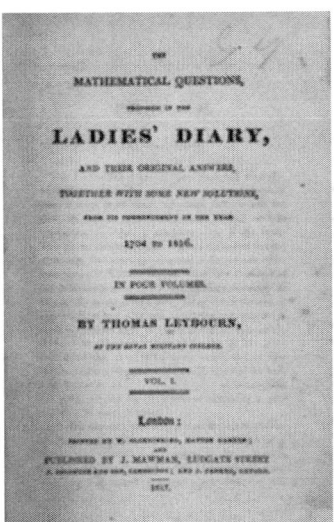

Figure 7.1. *Leybourn's compilation of The Ladies' Diary's mathematical problems and their solutions* provided a valuable resource for eighteenth- or nineteenth-century mathematical problem solvers. Image is supplied through Wikipedia Public Domain, Google-digitized via Hathi trust archive.

A Transition of Fluxions: The first question in *The Ladies' Diary* for which a solution was found using the method of fluxions was question 36, submitted by Barbara Sidway in the 1714 issue of the journal:

[Required] From a given cone to cut the greatest cylinder possible.

Four correct answers were submitted, one of which employed fluxions: the answer found to be 1/3 the altitude of the given cone.[13]

An example from a 1739 issue, reference number 210, submitted by a Mr. Richard Dunthouse, is more conceptually complex, requiring geometrically aided visualization:

Suppose a cask in the form of a middle frustrum of a hyperbolic spindle, whose length is 24 inches, bung diameter 30, head diameter 20, and traverse axis of the generating hyperbola 100 inches. Required its content in ale gallons?

Robert Heath, applying the theories of fluxions, determined the answer to be 49.893 gallons.[14]

Question 460 from the 1759 *Dia* is a dynamics-generated locus problem:

To determine the curve in which a body must move, so as to continue always at the invariable distance from another body moving uniformly in a right line; the velocity of the former body being also uniform, and exceeding that of the latter, in any given ratio.

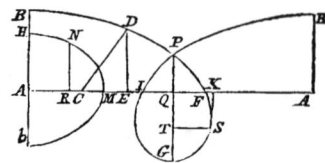

Using fluxions a Mr. O'Cavanah of Dublin determined the required path BDPFGIPBB described by body at D moving in conjunction with required body at C moving along line AA.[15] The middle of the eighteenth century saw an increase in problem solutions involving the use of fluxions. The *Diary*'s readership was by then more confident in the use of fluxions.

Question 710 from 1776, deals with a scientifically relevant topic of the time, physical properties of the earth:

> If, by experiments on large masses of matter, it is found that the mean density of the earth is $2\frac{1}{2}$ times the mean density at the surface, it is required to find the density of the matter at the centre of the earth, and the place or depth at which the density is equal to the said mean density, together with the law of the increasing of density all the way from the surface to the centre: supposing it is always reciprocal as some power of the difference between the whole diameter and the depth below the surface, and that the earth is truly spherical.

A Mr. "Mic." Taylor determined the density at the center of the earth to be 32 and the depth, $\frac{1}{3}$ the radius at the place of mean density.[16] In 1821 Charles Hutton published his findings on the mean density of the Earth for the Royal Society. His results were based on the observations of Nevil Markelyne obtained in 1774–1776.

A scientific controversy at this time concerned the dynamic geodesy of the Earth. Isaac Newton's theories of universal gravitation challenged the classical theories of the Earth being perfectly spherical. According to Newton, planetary attractions and interactions would result in a moving Earth to be an oblate spheroid, flattened at the poles and bulging at the equator. In contrast, Descartes's theories warranted the earth being a prolate spheroid: Newtonianism verses Cartesianism, continental theories verses those held in the British Isles. French expeditions were sent out in 1736 and 1737 to Lapland and Peru to obtain verifying measurements. Maclaurin, in his *A Treatise of Fluxions* (1772), supported the "oblate theory." In France, Jean d'Alembert and Pierre-Simon Laplace (1749–1827) mathematically investigated the situation. Eventually, Newton's theory was found to be correct.

In 1802, problem 1100 sent by a Mr. T. Hewett from London challenges the problem-solvers:

> Given the base and height of a cone, it is required to find the height of the greatest parabolic conoid which can be inscribed in the cone?

All four of the correct solutions obtained and listed were due to an application of fluxions providing the answers: $\frac{2}{3}$ the height of the cone or $(\frac{2}{3}\sqrt{2})$ the length of the cone's base.[17]

It appears that after 1835 the use of fluxions had been abandoned in favor of the continental system: differentiation and integration as conceived by Leibnitz. One of the last *Diary*'s examples that employed fluxions is number 1476 from 1827 where a Mr. Thomas Hornby of Wombleton proposes:

> Suppose ABCD to be a field in the form of a parallelogram, through which, if the line EF be drawn parallel to the ends, at the distance of c links from AD, it is found that the land on any part of the said line is worth, (n) shillings per acre; but decreases in value towards the end BC, inversely as the distance from AD. It is required to find the side of an equilateral triangle PQR, that will include as much land of the decreasing value as will be worth just (s) shillings; the vertex of the triangle being in the middle of the line EF, and its base parallel to the end BC.

The required answer found by fluxions is that the side of the triangle will be $2c(n-s)/s$ links.[18]

Thus, in examining the solution processes used in *The Ladies' Diary* over the period 1714–1841, fluxions come into vogue in the middle of the eighteenth century and later, by 1835, gave way to the techniques of differentiation and integration. Most of the problem situations encountered involved the determining of maximum or minimum values for a given unknown. During his tenure from 1745–1753, Robert Heath actively promoted the use of fluxions, an effort that was further strengthened by his critical references to Simpson's and Emerson's texts on the subject. This trend from fluxions to a more rigorous calculus paralleled that generally taking place in Britain at this time.

Geometry Takes a Turn. Coming into the eighteenth century, the analytical tool of English mathematics was algebra but from the problem-posing and solution methods available in the problems of *The Ladies' Diary*, it would seem that geometry surpassed algebra; John Playfair arrived at the same conclusion. While widely questioned as to its synthetic limitations, in the problem-solving event, geometry whether the main applied discipline or serving in a supplementary function, provided an analytic solution strategy. See chart in Figure 7.2.

Consider a selection of problems: In 1709, the readers of *Dia* were exposed to the following "grindstone problem," listed as number 9 in Leybourn's compilation:

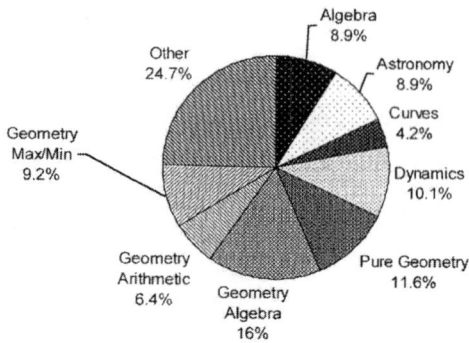

Figure 7.2. Survey of mathematical topics from Leybourn's Index, 1817 as tallied by Albree and Brown (2009) [**AB09**]. Courtesy of Joe Altree, Alburn University, with permission from Elsevier.

Seven men bought a grinding stone, the diameter of which was 5 feet, and they agreed that each should use it until he had ground away his share. What part of the diameter must each grind away?

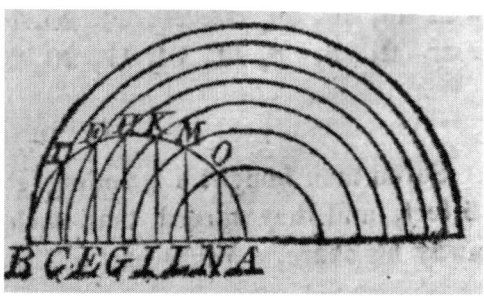

   While both the proposer and the solver of this problem remain anonymous, the solution process was provided in detail. The construction above has the radius of the given circle, AB, divided into seven equal parts and at the points of division erect perpendiculars meeting the circle constructed with the diameter AB?. Assisted by this diagram, the required shares are found, where the answers are rounded to two decimal places: 4.45; 4.84; 5.35; 6.08; 7.21; 9.39; and 22.68.[19]

   Question 313, proposed by Mr. Landen in 1749 is as follows:

There are 4 remarkable high trees growing in a straight hedgerow, the distance of the 1st and 2nd is 60 yards, of the 2nd and 3rd 40 yards, and of the 3rd and 4th 20 yards. Where must I stand to observe them so that the three intervals appear equal?

A Mr. Terey of Portsmouth supplied a correct answer aided by a construction:

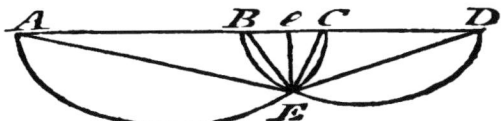

Where $AB = 60$ yards, $BC = 20$ yards, and $CD = 40$ yards, E marks the point of observation. Employing this diagram, the distances eC and eE are found to be 12 or 8 yards and 24 yards, respectively.[20]

The 1802 issue of the *Diary* contained problem 1090, submitted by a Mr. William Bewley from Hawkshead, requiring a solution obtained by construction:

Having the base, the sum of the sides and the vertical angle; to construct the triangle.

Three correct solutions were returned based on two different constructions. The construction demonstrated was provided by two problem solvers:

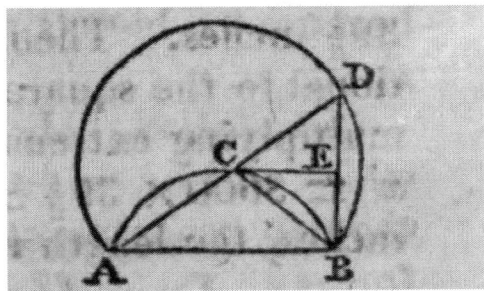

Here AB is the required base, the segment of circle AB contains an angle equal to half the given vertical angle and from point A, AD is drawn as the given sum of the sides. The required solution triangle is: CED or CBE, they are congruent.[21]

William Rutherford was a respected English mathematician who eventually became a teacher at RMA Woolwich. He is best remembered for his computation of the mathematical constant π to 208, digits, the best estimate at the time. In 1825, Rutherford submitted problem 1439 to the *Diary* that involved a mathematical situation that has now become known as "Napoleon's Theorem":

Describe equilateral triangles (the vertices being all outward or inward) upon the three sides of any triangle ABC: then the lines which join the centres of gravity of those three equilateral triangles will constitute an equilateral triangle. Required a demonstration.

The required demonstration was supplied by a Mr. Thomas Burn and a John Walker from West Boldon. Four other correct solutions were also acknowledged. In the diagram shown below, the required equilateral triangle is FED.[22]

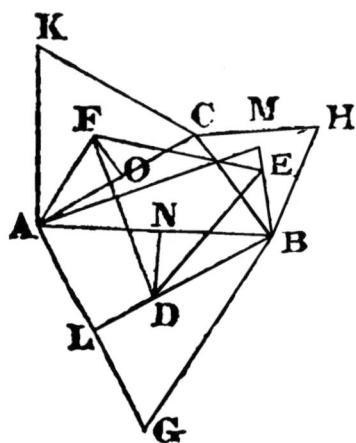

In the last issue of *The Ladies' Diary*, 1840, Mr. John Whitley offered the following challenging problem, 1670:

> If the lines bisecting the angles of a scalene triangle meet the opposite sides in three points, and each side of the triangles formed by joining these points be produced to meet a line drawn to its adjacent angle parallel to its opposite side, the three points of intersection will be in the one and the same straight line. Required a demonstration.

Mr. Whitley supplied a necessary demonstration in the 1841 edition of the *Diary*'s successor, *The Lady's and Gentleman's Diary*, pages 46 and 47.

In Leybourn's tallying of the *Diary*'s problems by mathematical discipline, geometry was easily the subject that occupied most problem-solving activity. Most geometrical procedures do not involve pure synthetic geometry but rather a mixture of investigative approaches: numerical, algebraic, considerations from solid and analytic geometries. Centuries of classical, mathematical conditioning are difficult to overcome. The editor of another contemporary periodical that also featured a section of mathematical exercises was besieged by his readers, requesting a lesser emphasis on geometry. While eager to satisfy this request, he noted that he was only publishing problems submitted by his correspondents and they were sending him geometry problems.

Analysis Asserts Itself. The concern of the nineteenth-century British mathematical community moving towards an analytic approach to mathematics can be interpreted in two ways: movement away from a synthetic conception of building up mathematics from a system of established rules and procedures, as ingrained in the study of geometry; or employing a strategy of beginning with the mathematical situation and seeking out the rules or applying and expanding the methods of the calculus [analysis], as developed on the Continent, using differential equations both ordinary and partial, the calculus of variations, applying mathematical series for approximations, etc. Both these trends of analytic strategy and analytic adoption are indeed evident in the progressing sequence of mathematical problems present in *The Ladies' Diary*.

Problem 1298 of 1815, the "Prize question," submitted by "Llerien" concerning the properties of a cycloid and "a path of quickest descent" is similar to investigations undertaken by Johann Bernoulli, Euler, and Lagrange:

> A cycloid being placed with the base horizontal, and vertex downward, it is required to assign the straight line of shortest descent from any point in the base to the curve.

Four correct solutions, including one offered by the proposer, were listed in the following edition of *Dia*.[23] The 1833 *Ladies' Diary* contains two adjacent problems: 1567 and 1568, that are analytic in their required solutions. The first, given by "Investigator," deals with complex numbers:

> When is $(a \pm \sqrt{-1})^{a' \pm b' \sqrt{-1}}$ a real quantity? Determine whether *all* functions $a \pm b\sqrt{-1}$ can be reduced to $A \pm B\sqrt{-1}$ or not. [This is a question, if not denied, by *Gergonne*.]

Note in this problem the use of the term "function." Reference is given to the French mathematician and logician Joseph Diez Gergonne (1771–1859). The following problem proposed by a "Petrarch," a frequent problem poser and solver, lies in the realm of differential geometry:

> If a tangent plane be drawn to a surface of the second order, cutting another surface of the second order, and where it cuts this latter surface tangent planes be drawn, the locus of the point where they all meet is also a surface of second order. It is proposed to demonstrate this.[24]

A lengthy proof and demonstration is supplied by a Mr. Mason of Scoulton, near Hingham, a persistent contributor to *Dia*.[25] Among the other successful solvers whose work is not shown is a "Miss L.L."[26] The anonymous lady had correctly answered several problems in this issue and was complimented by the editor who likened her to Donna Agensi and Mrs. Somerville. In a supplementary note, the editor called attention to the "admirable solution" provided for the converse situation published by two contemporary French mathematicians: Brianschon and Livet, in the *Journal of the Ecole Polytechnique*.[27]

Question 1637, 1838, submitted by Mr. John Cam of Torpenhow, Cumberland involves the calculus of variations:

Required the curve in which the integral $\int \frac{ds}{y}$ between given limits is a maximum.

The answer obtained by several correspondents is the equation for a circle with center at the origin and radius $\frac{1}{C^2}$, where C is an arbitrary constant.[28]

All such "analytic" problems had a large, ready audience of solution seekers. This phenomenon could be indicative of the increased knowledge of and an eagerness for Continental mathematical developments.

7.3 Further Mathematical Trends

A survey of *Dia*'s mathematical problems does reflect the calls for mathematical change then being expressed in Great Britain. Whether this emphasis was a reaction to or promotion of these movements is an open question but it exists. As John Leybourn noted in the "Preface" of his 1817 problem compilation:

The work may also be considered as exhibiting a view of the state of Mathematics in this country, and, on this account, must be valuable, as illustrative of the progress of the science in England during the last century.[29]

Throughout the publication life of *The Ladies' Diary*, the sequence of mathematical exercises was evolving: moving from rather standard recreational amusements to more challenging and creative settings and finally rather complex problems of a very specific applied and technical nature. The increasing intellectual and scientific shadow of the Industrial Revolution affected the direction of the *Diary*'s mathematical problem-solving efforts.

Also evident in the *Diary*'s exhibition of mathematical maturity was an increased referencing to existing mathematical literature and resources. In the early issues of the periodical, Emerson, and Simpson's works were often cited, later Charles Hutton's *Tables* (1804) and *Mathematical Dictionary* (1815) are called into service and during the later duration of the *Diary*, authors such as: Peter Barlow, David Gregory, John Bonnycastle, Samuel Vinces, Dionysius Larnder, and Henry Hamilton were referenced.

Occasionally, problems solved in the *Diary* were readily applicable to existing situations. One such problem, proposed as a challenging exercise, eventually found its way as an answer to a civic quandary. In the 1750 issue of the *Diary*, a question was printed that involved two bridges under which there were systems of water flow gates; dimensions were provided and it was asked how the distribution of waters should be taxed. Leybourn referenced this problem as number

331 (vol II, pp. 42–43). The "Scot-tax" allowed for the water was 100 pounds. "Britannicus" provided a correct solution by determining that the flow of water under each bridge was different and that the larger channel with more water should pay 85£1s 8d+ and the lesser channel 14£18s 3d+. In 1796, the East Sussex water administration experienced a similar situation involving their Fence and Penensey Bridges. *Dia*'s problem and its solution applied to their situation and helped resolve the issue.

The realm of the British Navy, its conquests and scientific exploits, combined with the facilities of improved communications, made the world, as a physical and geographical entity, better known to the English people. Some mathematically questioned its features: its shape, its density, and its gravitational attraction. The search for longitude prevailed on the national scene. British maritime activities combined with widespread trade involvement and a growing knowledge of exotic colonial possessions made English men and women global citizens, at least in a passive sense. The concern with navigational situations and computations exhibited in many of the *Diary*'s problems would seem to reinforce this impression. Some samples of this practice follow:

> Question 384 from 1754: Sailing due north, at the rate of 4 knots, in a current, a certain small island bore E.N.E. from us at a distance of 40 miles: After running 12 miles (by the log) it bore due east: having run 16 miles more, upon the same course, its bearing is then found to be S.E. To determine, from these observations, the direction and velocity of the current? (Answer: NLO 640 59', velocity 2.608 m/hr)[30]

> An 1806 question, 1153 has: A ship plying to windward $6\frac{1}{2}$ points of the wind (which is then N.N.E.) with her starboard tacks onboard 83 miles she then tacks and runs 62 miles farther. Required her difference of latitude made on both tacks? (Answer: the difference is 46.577)[31]

> "Philo-Nauticus" in the 1812 edition, in his question, 1252: How near must a ship sail to the wind, and what angle must her sails make to her course, so that she make the most way windward? Or, what angle must a ship's course make with the direction of the wind and what angle must her sails make with her course, so her motion in opposite of the wind, shall be the greatest? (Answer: Angle, the sail makes with the course, is 300; course makes with the wind, is 60^0)[32]

But the most persistent, and perhaps telling, trend in the mathematical problem sequences, was an increased concern with examples of applied mathematics: mechanics, dynamics, pneumatics, annuities, mensuration, and hydrodynamics. Many of these problems bore the status of "Prize Problem." One result of this movement was that the amateur mathematicians, Philomaths, and correspondents were replaced by more mathematically experienced profession-als: surveyors, engineers, military personnel, and varied practitioners; and aca-demics, schoolmasters, and university-trained clerics. Of course, this trend se-riously limited, in terms of interest and experience, the mathematical problem-solving challenges of the *Diary* undertaken by women. Perhaps prompted by the demands or concerns of the nation's increased industrial activities, the problems projected different conceptual outlooks:

> The Prize Question of 1784, 833, proposed by "Amacius": Two bodies A and B connected by a string or otherwise, at some invariable distance from each other, move, the one A along a given right line with a uniform [v]elocity, the other, B so that its velocity in the direction of the connect-ing line AB, may always be equal to that in a direction perpendicular to it. I demand the asymptote, equation, quadrature, and rectification of the path of B, its centre of curvature and the quadrature of the path of that centre. (Answer: Leydon, vol. III. pp. 137-139)

> Problem 1027 of 1797 concerns pneumatics: To determine the density of the air 1000 miles below the earth's surface; supposing a preformation be made to that depth, the diameter of the earth 7957 $\frac{3}{4}$ miles, the density of the air at the earth's surface to that of water, as $1\frac{1}{5}$ to 1000; and the pressure of the atmosphere on a square inch of the earth's surface $14\frac{3}{4}$ pounds, or 236 ounces. (Answer: 62583794... "to 71 places of figures" "—...amazingly great" [ounces/sq. inch ?])[33]

> In the last issue of *The Ladies' Diary*, 1840, question1673 was given by a Turkish Colonel Ameen Bey: Two balls are connected by an inexten-sible chain, one of which is placed on a smooth horizontal plane, and the other discharged from a cannon with a given velocity along a rectan-gular groove in the plane. It is required to determine their velocities at the instant the former enters the groove, the weight of the chain being neglected.

> (*The Lady's and Gentleman's Diary*, 1841, pp. 50-52 contains two correct answers, both obtained through a use of differential equations.)

In many situations for this latter type of applied problem, the "given condi-tions" constraints appear impractical or extreme and set as a hindrance to the problem-solver's endurance rather than an opportunity for a strengthening of mathematical understanding. "Mathematicus," a frequent correspondent with the *Diary* and a problem solver for several other periodicals, in an 1819 article, "In Defense of English Periodical Mathematical Works...," found in the *Philosoph-ical Magazine and Journal*, responds to a perceived slight to problem-solving-based journals attributed to a Mr. Henry Meikle. He challenges Mr. Meikle for the "depreciation which he applies to those periodicals in this country which are primarily devoted to mathematics" and suggests that he try his hand at solving a set of problems from the then just-issued edition of *The Ladies' Diary* for 1820. (This set of problems is given in Appendix B for the more mathematically adventurous reader.) Meikle did not "pick up the gauntlet." "Mathematicus" then went on further to defend the status of posing and solving mathematical problems as offered by contemporary periodicals:

> ...he [Meikel] would know that the mathematical sciences in this country owe the most solid obligations to those periodicals. He would know that while the managers of some learned societies that for many years have labored hard to stifle mathematical knowledge, those publications, by presenting a strong and varied stimulus to young investigators, have done as much if not more than even Cambridge and Oxford to keep it alive:[34]

As three examples of these periodicals, he cites: *The Ladies' Diary*, the *Gentleman's Diary*, and Leybourn's *Repository*. Furthermore, he lists a roster of British mathematicians who, he believes, were mathematically groomed by their participation in problem exercises offered in these periodicals. *The Philosophical Magazine* itself frequently considered mathematical problems along with a varied collection of articles and queries. For example, the "Defense" article, LXII,[35] was preceded by one on sea serpents and followed by Laplace's comments "On the Figure of the Earth." The mysterious Mr. Meikle defensively responded to "Mathematicus's" charges, protesting that he was misunderstood in his comments. He asserted and affirmed the value of the problem-solving exposure provided by the periodical and clarified that his objection/criticism was directed at the "ridiculous" problems often sent into these magazines.[36] In his diatribe, "Mathematicus" called attention to the "stifling" of mathematics in England.[37] The modern reader of these statements is probably left wondering, "Who is stifling mathematical knowledge?"—"Could it be the Royal Society?"

During the publishing life of *Dia*, approximately 2000 mathematical problems were posed, solved and sometimes discussed within its pages. When this number is extrapolated to account for the compilations made and the republishing of the *Diary*'s problems in other periodicals, the problem-solving exposure, experience, and impact on British society is impressive. John Playfair's 1808 comments as to the mathematical value of the *Diary*'s problems appear appropriate: yes, some problems were excellent, most mediocre and some ridiculous. Still, problem correspondents prided themselves on the number of correct solutions they found for *Dia*'s problems. In a memorial for the schoolmaster and mathematician Henry Clarke, it was noted as one of his lifetime achievements that he had correctly solved ten problems in *The Ladies' Diary*.[38] The trends exhibited by problem emphasis over the period do reflect the general mathematical movements taking place in England at this time. In this effort, the *Diary*'s mathematical problem-solving schemes were supportive and nurturing rather than innovative. Leybourn made a point of identifying this feature of the *Diary*'s problems in 1817.[39]

However, in 1795, Leybourn encouraged by the continued existence and popularity of *The Ladies' Diary*, had decided to publish his own journal devoted to mathematics. Entitled the *Mathematical and Philosophical Repository*, Volume I of this journal appeared in 1798 and was dedicated to Charles Hutton. A substantial publication, among its 426 pages could be found sections of mathematical questions and answers similar to those proposed in *The Ladies' Diary* but at a more difficult level of inquiry, of original essays, of "mathematical memoirs" extracted from notable published works, and information on recent English and continental mathematical publications. A reviewer would consider Leybourn's journal a natural extension of the *Diary*, one that supplied more focused and comprehensive information on mathematics; more of a professional journal rather than a recreational one. Contributions came from Woolwich: Hutton and Gregory strongly supported this new journal, but the majority of editorial work remained at Sandhurst where Leybourn was, eventually joined by his colleagues, Wallace and Ivory, in the production of the journal. After 1804, the periodical, now referred to as *Leybourn's Repository* [**Ley35**] appeared in a new series of six volumes in the years 1806, 1809, 1814, 1819, 1830, and 1835. During this period of great mathematical ferment in Great Britain many of the nation's most eminent mathematicians contributed to the *Repository*, besides the directly contributing editors. Included in this list were Peter Barlow, Benjamin Gompertz, John Herschel, William Horner, John Peacock, and Mary Somerville. *Leybourn's Repository* was extremely influential in the reforms taking place. Niccolo Guicciardini in his study of British calculus changes taking place at this time credits the first three volumes of Leybourn's new series of the *Repository* as: "...one of the most important works in the reform of British calculus."[40]

Whether the knowledge obtained by periodical-offered problem-solving exercises exceeded the mathematical training provided by Cambridge and Oxford, as claimed by "Mathematicus," seems doubtful but they certainly broadened it. *Dia*'s mathematical problems mirrored the industrial and scientific concerns of the times. In its popularity and utility, *The Ladies' Diary* certainly weathered the "Winds of Change" but it also fostered the emergence of an advanced offspring, *The Mathematical and Philosophical Repository*, a predecessor to the modern mathematical research journal.

8

Women and the *Diary*

Did the *Diary* really serve the needs of women?

When at first the modern readers' attention was brought to *The Ladies' Diary* by the works of Perl (1977, 1979) and Costa (2000) [**Cos00**], it was evidence in the ongoing debate of women in mathematics. "Could they do mathematics?" "Should they do mathematics?" Both the testimony of the *Diary*'s existence and content provided dramatic and affirmative responses to these questions. While the cultural biases against "women and mathematics" are now identified, better understood and mostly mitigated, some lingering questions remain as to the place and function of *The Ladies' Diary* within its society.

8.1 Was The *Ladies' Diary* truly a Ladies' Periodical?

Certainly John Tipper conceived, developed, and carefully nurtured his almanac for women. He would employ "no unchaste words, with harsh offensive sounds."[1] These efforts would seem to be more than a mere simple business accommodation; Tipper, as a teacher, realized the neglected intellectual potential of women and their need for an informative and educational venue. In his correspondence with his confidant Wanley,[2] he made his intentions quite clear to serve a female readership: he distanced himself from the rhetoric of burlesque almanacs so that no content would "raise a blush." Noting the increase of women's growing interest in science, his instructions and lessons on science would be modern; he would avoid politics; discourse was to be pleasant and genteel; enigmas, and later mathematical problems, would be given in verse. While attempting to accommodate his female audience, Tipper also held biases that would, initially, be reflected in his journal: The *Diary*'s calendar renderings would have "a great variety of particulars all at length because few women make reflections or, are able to deduce consequences from premises."[3] His female correspondents would soon prove his assertion wrong.

At the time of its appearance, *The Ladies' Diary* was the only periodical on the market designated and compiled specifically for women. Tipper developed a congenial relationship with his readers by encouraging an interaction through correspondence. Each of his *Diaries* opened with a personal, welcoming introduction to his lady readers. His female audience became actively engaged, appreciating the opportunity to have their opinions heard and, in many cases, acted upon. Tipper's prompting them to solve his published enigmas resulted in a request for even more difficult puzzles. Similarly, when mathematical questions were introduced into the *Diary*, readers requested that the practice be continued and expanded. Tipper censored such questions for their appropriateness for ladies. Thomas Leybourn in reviewing this mathematics in 1817 commented negatively on its quality.[4] There is little question that under John Tipper's editorship, *The Ladies' Diary*, in intention and "ownership", belonged to women: he designed it for women, formally dedicated it to them, and reacted to their suggestions. Myra Reynolds in her 1920 study, *The Learned Lady in England*, 1650–1760, noted that *The Ladies' Diary* offered English women "a genuine delight in personal mental activity" and that in the period 1703–1726 no other agency "offered to women such an intellectual opportunity."[5] *The Ladies' Diary* projected an aesthetic rationalism. But with the string of successive male editors and changing market conditions, the ladies' ownership and the editorial direction of the periodical would come into question.

It would be the excesses of one editor, John Heath, and the mores of nineteenth-century England that would alter the appeal of the *Diary* to women and, in turn, increase a male audience for the mathematical exercises. Socially conscious ladies would be discouraged from engaging *The Ladies' Diary* as a result of Heath's use of inappropriate, bawdy references and the public quarrels and controversies surrounding the editor. Most heinous was his muted but published ridicule of a young Lady's (Sophia Western) mathematical solution.[6] Thus, although the almanac still bore the title *Ladies' Diary* and much of the material— questions and enigmas—were addressed to "Ladies," it was only nominally a ladies' journal. Ladies no longer possessed the ownership of the periodical in the sense that they did not actively control the direction nor intent of its features.

After John Heath, the following editors attempted to restore the integrity and reputation of *The Ladies' Diary*. They improved the setting of mathematics problems and continued to solicit female readers, but the aura of intellectual freedom and the opportunities for women's self expression instilled by John

Tipper and nurtured by Henry Beighton together with his wife Elizabeth were gone. For women, the air of liberation, expression, and participation was diminished. Mathematical encounters became threatening rather than entertaining. No more "friendly chats" by the editor directed to the readership would preface content presentation. *The Ladies' Diary* now became a recreational magazine for both sexes. For the rest of its existence, it would remain only in name a "Ladies' " periodical but continued to promote puzzles and mathematical exercises that appealed to and were consumed by both sexes.

8.2 The *Diary* as a Vehicle for Female Mathematical Expression

Researcher Shelley Costa made some tentative associations for a few female problem-solvers. Her findings are valuable and will be reviewed.[7]

The young maidens in the complexion of Jane Austen's heroines who challenged the conventions of the times to demonstrate (and perhaps flaunt) their mathematical abilities were young in age, early teens to beginning twenties: Mary Wright in 1710 when she won the Prize Question was eighteen years old, her sister Anna Hannah was just sixteen when she captured the Prize in 1711;[8] Miss Nancy Mason of Clapham was nineteen when she began her six-year-long, fruitful, problem-correspondence with the *Diary* (1792–1798)[9]; the elusive Miss T. S___e (1758), with a dowry value of 4410 pounds, was twenty-one.[10] All of these ladies were categorized at the time as "being of marriageable age." The Wright sisters were from Cheshire;[11] their father, the Rev. Matthew Wright, had obtained a degree from Oxford University and had probably home-tutored his girls in mathematics. A cousin, Thomas Wright, who lived on a nearby estate, also participated in solving *Dia*'s problems. Perhaps, the three young people formed a "study group" to jointly discuss and tackle the problems. It seems most likely that the *Diary*-centered *Philomaths* would be intellectually and socially attracted to each other. Sylvia Harrop, local historian for Clapham, Yorkshire, was frustrated in her efforts to better identify the Nancy Mason acknowledged in the *Diary*'s pages for her "ingenious answers." Available official residential records for the period do not mention the "Mason" name. Harrop speculates that Nancy might have been employed in the area as a governess.[12] Whatever her station, Miss Mason conceived imaginative mathematical situations as demonstrated in the context of a 1796 problem she submitted to the *Diary*:

> A father dying left a square field to be divided among his five sons, the field containing just 30 acres, and to be divided in such a manner that the oldest son may have 8 acres, the second son 7, the third son 6, the fourth son 5, and the fifth, or youngest son, 4 acres. Now the fences are to be so made that the oldest son's share shall be a narrow piece of equal breadth

all round the field, leaving the remaining four shares in the form of a square; and in like manner for each of the other shares, leaving always the remainders in the form of squares, one within another, till the share of the youngest be the innermost square of all, equal to 4 acres. It is required to divide the field geometrically, and to calculate the sizes of all the fences?[13]

Another Miss, age unknown, who displayed her mathematical talent by answering eight questions in the period 1809–1816 was a Mary Weston, from Gainsborough. As previously mentioned the "Miss L.L." won the editor's accolades for her problem solving in 1815.

But in the early days of the *Diary*, it was married ladies who demonstrated an outstanding, versatile, mathematical ability and a feminine pride in their rebukes to male challengers. In particular among this cadre were Mrs. Barbara Sidway and Mrs. Mary Nelson. Mrs. Sidway frequently accompanied her correct answers with an insightful extension of the problem—"What if?"[14] Mary Nelson sometimes included a telling retort to a problem's male proposer. For example, the Prize question for 1714 concerned the gauging of a spherical cask which contained several liquid-displacing foreign objects. Besides displaying her knowledge of physics, Nelson prefaced her correct reply, given in verse, with some words of consolation:

Alas poor man! you're in a sad condition;

Who can refuse to answer your petition?

If th' ladies grant you not relief with speed,

You may at last be so reduc'd indeed

To ride like Hudibras, on such a steed.

But since you are in such a pitious case,

For once, I'll save your credit and your place:

Here underneath I've answer'd all your queries,

Which gives me one fair chance to win the diaries[15]

The following year's Prize question concerned navigation and was judged by the editor to be both complex and difficult. He noted an abundance of incorrect answers as evidence for his claim. The question, by its subject matter, was hardly considered suitable for a lady. Despite this challenge, Mary Nelson, along with nine other male correspondents, provided the correct answer and commented with a lady's justifiable reaction:

Master Tar I protest I'm not pleased with your query,

You've sent at this time to be solved in your Diary,

Methinks it looks like an ill-natur'd demand,

To offer to puzzle us ladies at land;

Tho' I'm apt to believe, if a trial should be

That you'd puzzle us worse if you had us at sea;

But methinks I can guess how you sparks are inclin'd,

You are always best pleas'd, when the ladies prove kind;

Then to shew that we maids of this brave British nation,

Can act here at land, or at sea on occasion,

To each of your queries, I've sent you an answer,

And all of them right, then deny't if you can, sir.

(Here follow'd the answer.)[16]

She was a competent and spirited mathematical competitor. Both women exhibited the same *Diary* participation pattern: answering a few questions in a series, then two prize questions—Barbara Sidwell: 1712 and 1716; Mary Nelson: 1714 and 1715, before disappearing from the *Diary* scene. Perhaps they were testing their skills at the problem challenge and, as they succeeded, proving themselves, they quit corresponding with the journal.

During the nineteenth century, the apparent lack of female mathematical problem solvers in *The Ladies' Diary* was not necessarily indicative of their declining interest in the subject, for now they could be satisfied elsewhere. Following the *Diary*'s success, a whole series of alternate mathematical problem-solving venues sprang up: other periodicals, newspaper columns and organized female mathematical study groups or clubs. The witness of the *Leeds Correspondent* in the first quarter of the nineteenth century indicates the ongoing prevalence and vitality of British women doing mathematics.[17]

8.3 The Ladies' Statements

Previous major studies of *The Ladies' Diary* such as those undertaken by Perl, Costa, and Meigon have relied on a "body count method" of female participation in the *Diary*: number of female respondents to enigmas or mathematics problems.[18] Such an evaluation strategy is misleading. Perhaps the tally is true that in the period 1709–1713, apparently 78 men responded to the mathematical questions as compared to 9 women and that by the later period 1744–1753, the numbers had changed to 165 versus 14. For answering enigmas, the gender differences are less disparate: 1709–1713, 51 versus 40 and for 1744–1753, 225 and 106.[19] Yes, there is little doubt that throughout the existence of the *Diary*, more men responded to its features than women, but the available reading audience of men at this time was greater than that of women! Estimates of English literacy rates in 1714 indicate that 45 percent of men could read and write as compared to 25 percent for women, almost twice as many men as women. By 1750, these percentages advanced to 60 percent and 40 percent or three to two. Males bragged and sought exposure. They "advertised" their services via the Journal, thus their certification often included a title such as: "schoolmaster," "Rev." or "surveyor." Furthermore, the popular use of pseudonyms cloud the actual participation tallies. It would seem that in the social/cultural climate where a woman's visible correspondence with a magazine, especially on a subject such as mathematics, would be deemed highly inappropriate, culturally deviant, and a possible affront to her husband, it would warrant the use of a pseudonym, often that of a male. So the numbers themselves are misleading. What is important is that women, for a short period of time, ostensibly had their own designated journal and that they shaped it in the manner that they did, towards mathematics. We don't know how many actually read the journal and passed it on to their friends or fervently discussed its contents among themselves, or tried a solution and failed. Obviously, many more! As previously recognized, there had been predecessor journals intended for women but these journals were designed to "improve" the female, remove her assumed deficiencies, improve her morals, and strengthen her, to better mold her to an image conceived and approved by men.[20] As an example of the female mentality these periodicals preyed upon, consider one example from a "Question and Answer" column in 1691 where the woman inquired: "Is it proper for a woman to be learned?"[21]

Whereas *The Ladies' Diary* offered scientific instruction, appropriate entertainment, and the opportunity to interact with the editor and other readers, both female and male, it was the ladies who requested more demanding enigmas and the availability of mathematics questions. At this period of history, there was no other popular periodical containing mathematical exercises or problem solving for its readers. *The Ladies' Diary* was a unique intellectual statement for its times and initially, a resource by and for women. Through their actions, the ladies made several profound statements for the time.

- Firstly, mathematics and its computational problem solving were important.

- Secondly, they, as women, wanted to do mathematics.

- Thirdly, they could do mathematic quite well, even beyond their societal-set horizons of arithmetic.

- Lastly, by their continued domination of unraveling of enigmas, women were good problem solvers.

In total, this was a formal, public declaration of a woman's intellectual independence and abilities.This is its crowning glory, the real magic of the appearance and existence of *The Ladies' Diary*!

8.4 Prominence and Recognition

While British women's involvement with the mathematical exercises of *The Ladies' Diary* attested to their mathematical interests and abilities, it also demonstrated their resolve to obtain new knowledge of which they had been deprived. Remember that a young English lady of this era might, at best, have been exposed to a study of simple arithmetic. Formal training in algebra, geometry, trigonometry, "philosophy" or the physics of kinetics and dynamics, and the theory of fluxions were "male" prerogatives. Ladies, if they wished to pursue such advanced studies, relied on their individual determination and resourcefulness. See Figure 8.1.

Figure 8.1. A Hodgson cartoon of the period. While the young girl's father attempts to teach her geometry, she appears distracted by a suitor hiding behind the globe. A viewer can arrive at several conclusions. Credit: A middle-aged man giving a geometry lesson to young woman, a young man [her suitor?] hides behind a large globe. Coloured lithograph. Wellcome Collection. Attribution 4.0 International (CC BY 4.0)

Most certainly, the wives and widows of tradesmen and shopkeepers in eighteenth- and nineteenth-century England knew and practiced simple arithmetic and helped to maintain the family business. And the female mathematical problem posers and solvers of *The Ladies' Diary* demonstrated that able and apt female *Philomaths* existed at that time. The cadre of the *Diary*'s early problem solvers including Nancy Mason, the Wright sisters, the elusive "Adrastia," and Barbara Sidwell support this fact. Even when the more mathematically demanding *Mathematical Repository* appeared, a Susan May distinguished herself among the mathematical problem solvers, 1804–1814. In the 1811 issue of the new *Repository*, Mary Somerville answered a difficult mathematical problem involving Diophantine equations, winning a silver medal for the effort. It would be a worthy and revealing task to survey the variety of then existing mathematical problem-solving periodicals to determine the wider female participation.

Mary Somerville (1780–1872) was already a respected mathematician and scientist in the English community.[22] Her ultimate scientific contribution would be the English translation and commentary of Pierre Simon Laplace's *Mécanique celeste as Mechanics of the Heavens*, 1831. It should be noted, as fitting her prescribed status as a married woman at this time, she signed herself author as "Mrs." Mary Somerville. Lord Byron's daughter, Ada, later to become the Baroness of Lovelace (1815–1852), studied mathematics under Mary Somerville as well as with the mathematician Augustus De Morgan [**DM47**]. Ada Lovelace then went on to become an assistant to Charles Babbage (1791–1852) in his "Analytic Engine" experiments [**Bab30**]. After Cambridge opened its colleges to women in 1848, several ladies distinguished themselves in mathematics: Charlotte Angas Scott placed eighth wrangler on the Tripos in 1880; Hertha Marks completed the Tripos in the following year; and Sophie Bryant in 1882 obtained a BSc from the University of London and would complete a DSc in 1884; a "Miss Burstall of Girton College" earned a Senior Optimes standing. In 1890, Philippa Fawcett of Newnham obtained the unique distinction and associated fame of exceeding Senior Wrangler rank status in her mathematical performance.[23] At the time, Miss Fawcett's mathematical status warranted mention in North American newspapers and even the *Times of India*. Female ability in mathematics and problem solving in Great Britain has always existed and if allowed, would flourish and blossom.

9

The Ladies' Diary, a Noteworthy Heritage

So what did we learn about the *Diary*'s effects and societal impact?

9.1 A Wider Influence

In examining the life and influence of *The Ladies' Diary*, we have encountered many accolades to the journal's credit. Certainly the testimonies of John Playfair and Thomas Kirkman, as given in Chapter 2, were sincere at the time. *Dia*'s influence was wide and great. Its subscribers were found all over the British Isles, in Europe and America. "Mathematicus" in his 1819 "Defense of English Periodical Mathematical Works", noted how "able philosophers" from France, Germany, Denmark and other countries were so impressed by *The Ladies' Diary* that they brought home copies of the journal to emulate.[1] The eminent British astronomer and mathematician William Wales, when he went on an expedition to Hudson Bay in 1768 to observe the transit of Venus, had copies of the *Diary* brought to him by ship. In a letter home, he commented that they "make a feast for the greatest mathematical epicure in the world."[2] Wales, himself, had been a contributor to the *Diary*. Even non-subscribers of the periodical were made aware of *Dia*'s existence: the journal's popularity amongst women was mentioned in the novel, the *Histories of Lady Francis S___ and Lady Caroline S___*, 1763 and Hannah Cowley's play, "The Belle's Stratagene, A Comedy."[3] Mary Wollstonecraft, in her *Vindication of the Rights of Women,* 1792 [**Wol92**], credited *The Ladies' Diary* as the:

> ...earliest, longest-lived and most successful of all periodicals aimed at women.

While such reports certainly support the high esteem for *The Ladies' Diary* held by many individuals, stronger measures of its impact exist. Firstly, it survived in a highly competitive publishing field for nearly a century and a half, an impressive record for the times. Secondly, imitation has been singled out as the greatest compliment an individual or institution can accrue. *The Ladies' Diary* was frequently imitated and collections of its contents, particularly mathematical problems, culled for other supplementary periodicals and journals. The "Family Tree" chart given below attempts to demonstrate and summarize some of these efforts.

A Family Tree of Offspring

Delights for the Ingenious, 1711

The Gentleman's Diary or Mathematical Repository, 1741–1840

A Miscellany of Mathematical Problems, 1743

The Mathematician, 1745–1750

Mathematical Miscellany, 1747–1755

The Palladium or an Appendix to The Ladies' Diary, 1748–1779

Mathematical Exercises, 1750–1753

The Ladies' Philosopher, 1752–1754

The Ladies' Chronologer, 1754

Darian Miscellany, 1764–1776

The Darian Repository, 1774

Ladies and Gentlemens Diary or Royal Amanack, 1775-1786*

A Supplement to The Ladies' Diary, 1788–1826

The Gentleman's Mathematical Companion, 1798–1826

A Companion or Supplement to The Ladies' Diary, 1791

The Diary Companion, 1792–1806

Mathematical and Philosophical Repository, 1795–1803

Mathematical Repository, New Series, 1804–1835

Mathematical Questions in The Ladies' Diary, 1817

The Northumbrian Mirror: or Young Student's Literary Mathematical Companion, forming an Introduction to The Ladies' Diary, 1837–1841

The Lady's and Gentleman's Diary, 1841–1871

---

*In 1775, Thomas Carnan, a London printer and bookseller, broke the monopoly on almanac publishing held by the Company of Stationers. From 1750 onwards he had been producing imitations of *The Ladies' Diary*. Now, freed from legal constrictions, he published a more direct version of the *Diary*, both in format and content, entitled *The Ladies and Gentlemans Diary or Royal Amanack*. To make matters even worse, in 1780 Carnan dropped the word "Gentlemans" from his title, the word "Lady's" was changed to "Ladies'" and then there were two Ladies' Diaries on the British market. For more details, see Kon, 2012 [**Kon12**]. By 1777, Carnan controlled the publication of thirty-one almanacs of which eight were designed for ladies. When he died in 1788, the ownership of all his diaries reverted to the Company of Stationers [**Bla61**]

The listings include only "first-generation" descendants, that is, publications, which most directly replicated the *Diary*'s features. Still many other publications such as newspaper columns dealing with mathematics problems, owe their inspiration to the *Diary*'s popular reception.

The Ladies' Diary was not an academic or scientific research journal but it was an important precursor to the existence of such journals. Writing in the 1880 issue of *Nature*, the eminent Victorian mathematician and astronomer and editor of the Cambridge publications, the *Quarterly Journal of Mathematics,* and later, *The Messenger of Mathematics*, James W.L. Glaisher (1809–1903) [**Gla80**] divided the development of British mathematical periodicals into three consecutive categories:

(1) Periodicals devoted to puzzles and requiring the solutions to provided, mathematical, problems. Correct solutions would be given in the following issue and prizes awarded for correct answers. Some extended commentary might accompany the published solutions.

(2) A periodical with two distinct sections. One section would be devoted to original papers; the other to problems, posed or solved.

(3) A true scientific research journal containing original, refereed papers.[4]

Clearly, Glaisher had *The Ladies' Diary* in mind when he formed the first category; however, under the editorship of Olinthus Gregory, it began to move into the second category with added appendices containing articles and papers. The *Diary* foreshadowed proper research journals.

The beginnings of the nineteenth-century saw three forums of mathematical research develop on the Continent: Joseph Gergonne's *Annales des mathématiques pures et appliquées*, 1810, Paris; August Leopold Crelle's *Journal für die reine und angewandte Mathematik*, 1826, Berlin and Joseph Liouville's *Journal de mathématiques pures et appliquées*, 1836, Paris. Of these three, Crelle's *Journal* was the most active and influential. It was not until 1837 that Great Britain established such a formal vehicle for mathematical information and interaction in the form of *The Cambridge Mathematical Journal*.

9.2 Offspring in America

Lady Di also inspired similar periodicals to be started in America. George Baron (1769–1812), an immigrant from England, became a mathematics instructor at the new American Military Academy at West Point. In 1804, Baron began to publish a mathematical periodical directly fashioned after *The Ladies' Diary*. He called it the *Mathematical Correspondent*. As a young man in England, he had solved several problems in the *Diary* and felt that a journal of its kind would benefit the people of his new country. Soon after initiating this project, Baron died. Editorship of the fledgling journal passed into the hands of Robert Adrian (1775–1843), also an expatriate from England and a self-taught mathematician. Baron had managed the *Correspondent* poorly and its new editor could not save it. It faltered; but in 1808 Adrain began a new journal, *The Analyst or Mathematical Museum* which also reflected *Dia* in its purpose and composition. Readers found the enclosed mathematics too difficult and this journal lasted until 1814. A schoolmaster in New York City, Melatiah Nash (ca. 1768–1830), published another periodical influenced by *The Ladies' Diary*.[5] Started in 1820, he called his publication the *Ladies' and Gentlemen's Diary*. It only lasted two years. Undiscouraged, Robert Adrain attempted a new journal, *The Mathematical Diary*. This, his last publishing effort, ran until 1832.[6] In 1836, still another immigrant from England and a frequent correspondent with *The Ladies' Diary*, Charles Gill (1805–1855) who is recognized as America's first actuary, began his journal, *The Mathematical Miscellany*. It was to last three years.[7]

Thus, all the early mathematical periodicals began in America owe their format and content to their British cousin, *The Ladies' Diary*. In 1743, Benjamin Franklin founded The American Philosophical Society in Philadelphia, Pennsylvania. Franklin's efforts were directed at forming a center for scholarly learning. Today, among the Society's historical collection is found a substantial holding of *The Ladies' Diary*, 1714–1752. Finally, over a century later in 1878, would the United States be able to boast of a true mathematics research journal, the *American Journal of Mathematics*.

9.3 A Conclusion

Indeed, *The Ladies' Diary* has many stories to tell and insights to provide. Our narrative has taken us through movements: a popular acceptance and use of mathematics by common people; a demonstration of women breaking with constricting traditions and seeking a rightful place in their society, expressing their intellectual interests and ability; the evolution of popular mathematical and scientific publishing; and the changing mathematical priorities resulting from economic and technological pressures. This periodical offered much to its readers—it supplied information, promoted a scientific spirit, provided intellectual challenges in the form of enigmas and mathematical problems, encouraged and developed problem-solving skills in readers. These skills were both literary and scientific. In examining *Dia*'s impact, we have also traced the evolution of the obvious reading audience in which the momentum of mathematical participation moved from *Philomath*, talented ladies, to trained school masters and clergyman, to engineers, technologists, military men, and more formal academics. But knowledge of the *Diary* and its influence reached and affected many more individuals. Perhaps they learned of a question and repeated it and/or discussed it with colleagues or friends. Even such minor interactions promoted mathematical awareness. How many persons tried solving some of the enigmas or problems and gained experience at problem solving even though they may not have succeeded? How many people succeeded and did not seek recognition for their work? Certainly, these populations were greater than those that were recorded in the pages of the periodical. Similarly, large populations of problem solvers were served by the summaries and compilations that were made and published by such men as Hutton and Leybourn. Both works remain valuable resources into the mathematical thought and practices of the early Victorian Period.

In this examination of *The Ladies' Diary*, we have come across mathematicians who have distinguished themselves by their participation in the periodical's challenges or by their association with it, such as Thomas Simpson, Charles Hutton, Olinthus Gregory, John Landen, John Playfair, and Robert Adrain. This listing could be extended to a much greater inclusion, considering less obvious problem solvers whose careers and accomplishments did not necessarily focus on mathematics. No doubt there were several, both males and females, who did benefit scientifically and intellectually from their contact with the *Diary* and were not highlighted as mathematicians within its history.

One such individual worthy of further note is John Dalton (1766–1844), one of Great Britain's most eminent scientists. Dalton, recognized as a chemist, physicist, and meteorologist and having established the Atomic Scale, is primarily remembered as the "Father of the Modern Atomic Theory."[8] Dalton was from a Quaker family of weavers and, being a Quaker, was denied university entrance. In his higher studies, he was tutored by a blind philosopher and relied on his own personal efforts. *The Ladies' Diary* played a fundamental role in this educational quest. At age thirteen, he copied verbatim an issue of the *Diary* to study, and throughout his early career was a frequent contributor to the mathematical problem section. Thomas Leybourn credits him with submitting sixteen correct solutions, including one prize solution, to the periodical—an impressive record. Karen Zwer, in writing on Dalton's scientific career, believes he approached his research problems as puzzles: they were "puzzle-centered."[9] Perhaps, his problem experiences from the *Diary* shaped this strategy. Upon Dalton's death, his library was found to contain a large collection of *The Ladies' Diary*. In 1779, the "Prize Question" was answered by the noted astronomer William Herschell (1738–1822). He, as thousands of others, was attracted to its problem challenges. Leybourn described the journal as: "...a curious, and valuable monument of the mathematical genius of the English nation."[10] Certainly, it was!

Despite the varied audiences that contributed and benefited from the existence of *The Ladies' Diary or Woman's Almanack*, it was the initial female correspondents that demonstrated women's interest and ability in mathematics and who asserted themselves in the field of problem solving that allows them to stand out as true intellectual and social innovators. They should be remembered as establishing the fact that women could do and would do mathematics.[11] Furthermore, they were competent and able problem solvers in the broader sense of understanding premises and reaching viable conclusions. This latter demonstration was expressed by that continual involvement in the solution of enigmas. Throughout its existence, *The Ladies' Diary* was a declaration for and of the potential of women.

From the Golden Age of Queen Elizabeth I with the ascension of England as a naval power through the reign of Victoria, and the beginnings of the industrial revolution, the popularization of mathematics in Great Britain continued to grow. It was an egalitarian educational movement designed mainly for the common people, the workers. Practitioners and reckoning masters promoted the use and learning of mathematics, stressing its utility. Merchants and trade houses supported this effort in order to maintain England's dominance of the seas thus insuring their economic wellbeing. A vital element in this evolution of new understanding was the appearance of a simple periodical, *The Ladies' Diary or Woman's Almanack*. It was initially conceived as a recreational and scientific resource for women but it evolved to become an exemplar and stimulus for mathematical problem solving. The changing content and increased complexity of the journal's problems reflected and supported the mathematical movements and reforms that Great Britain was experiencing. In turn, the mathematical sophistication and involvement of the editors grew accordingly: from that of a well-meaning schoolmaster and amateur mathematician, John Tipper, to a nationally respected mathematician, Charles Hutton, professor at the Royal Military Academy, Woolwich, and Fellow of the Royal Society. Thus, the three forces isolated earlier in our investigation, in chapter 2, have been examined in some detail and their influence on the existence and development of *The Ladies' Diary* better understood:[12]

(1) The course and intellectual direction the *Diary* took was a direct result of its editors' backgrounds, interests, and experiences. All were male, with the exception of Elizabeth Beighton's brief tenure. Professional mathematicians and avid problem solvers occupied the span of editorship from 1754 to 1840, the majority of *Dia*'s existence.

(2) At the onset of the journal's publication, it was women who requested and established a mathematical problem-solving section. Furthermore, they went on to demonstrate their interest and ability in the challenges of mathematical problem solving. Their participation in this feature waned as the level of mathematics moved beyond their limited education and training. They did, however, maintain their continued eminence in the posing and solving of word enigmas, attesting to their ability as problem solvers in a more general sense. Throughout the *Diary*'s existence, it marked and promoted an intellectual era for women by retaining the words "Ladies" and "Womens" in its title.

(3) Yes indeed, *The Ladies' Diary or Woman's Almanack*, in its contents, very much reflected and supported the mathematical movements and reforms of the period. How well it influenced them remains an open question.

For over a century, 1704–1840, *The Ladies' Diary* served as a dynamic forum for mathematics learning, teaching, and understanding. It promoted the doing of mathematics. *Dia* remains a milestone in the development of British mathematics.

Epilogue

What more can we learn from *The Ladies' Diary*?

At the beginning of this project, I submitted an outline of my research objectives to several reviewers. Much to my surprise, one reviewer commented: "Why are you bothering with *The Ladies' Diary*? It's all been done." These words encouraged rather than discouraged my efforts, as they indicated how little even supposedly knowledgeable people knew about *The Ladies' Diary*. Yes, a few articles had been written about this periodical but almost all of them focused on its initial years of publication when it was more completely a woman's magazine. Their emphasis was on feminist accomplishments. The influence of this periodical lasted over a century and a half during a period of English social and intellectual flux. How much was *The Ladies' Diary* a participant in this change? Did it contribute to the change? Did it reflect the change? What did the *Diary* tell us about the mathematical movements of this time, the educational changes? So many open questions seemed obvious to me! Perusing several issues of the journal, I found a cornucopia of concepts and unanswered issues that attracted my attention and curiosity: the evolving situational complexity and mathematical emphasis of problems, the limited participation of the female subscribers to later mathematical problem challenges, and the popularity of the enigmas as puzzles. The above presentation attempted to clarify some of these issues. While in the process of this investigation, new questions arose such as: "Did the female problem solvers who became disenchanted with *The Ladies' Diary* turn to other periodicals of the time to exercise their mathematical interests? Did they continue doing mathematics from the *Diary*, discreetly employing a pseudonym? How pervasive was the scope of women's mathematical study clubs? Did women openly participate in mathematical problem-solving challenges offered in the newspapers? Are there formal, extant female statements of this period expressing women's interest in mathematics? Did the content of the *Diary*'s enigmas change over time, reflecting societal priorities? How popular and active were private female arithmetic teachers in England prior to and during the lifespan of *The Ladies' Diary*?" During the textual discussion, I made speculative statements. These were intended to be "open-ended," allowing for further clarification and encouraging future research. I hope that some readers will pursue and resolve these further issues. Similarly, I hope that avid problem solvers will seek out some of the original resources given in the bibliography and attempt their problem challenges. "Can a contemporary, mathematically literate individual satisfy the challenges posed by Victorian Euclidean geometry exercises of a hundred years ago?" It would be an interesting experiment to obtain solutions for eighteenth- and nineteenth-century problems using modern mathematical techniques and compare results for efficiency. For those who have access to the Internet, the Hathi Trust Archive contains copies of *The Ladies' Diary* from 1763–1840, as well as issues of Hutton's and Leybourn's problem compilations. These works are readily available. *Dia* still has much to offer!

Appendix A

Selected Word Puzzles from *The Ladies' Diary*

I. (1763)

V. Enigma 455

I equally belong to young and old,
The coy, the fond, the modest, and the bold.
In woman's composition I am found,
But not in man's; and yet I wander round
The mossy fountain, and the gloomy groves,
With gentle Ovid and the jocund loves.
Old Homer too and Horace own my power;
I sport w.th Milton in the noon-tide bower:
Chiefly to me you owe the poet's song,
When the smooth verse more gently flows along
Than Chrystal streams that o'er the pebbles stray,
And, softly gliding, serpentine their way;
Yet you may find me when the ocean roars,
And monstrous billows lash the sounding shores.
Always in motion, yet I never stir;
In oaks I dwell, but shun the beech and fir:
In history too, both sacred and prophane,
You'll know me by my shape, my voice, my name,
Inconstant as the moon, yet still the same.
To find out my beginning, and less where,
Or when I end——that task is too severe.
But hold! enough! ladies, I make no doubt,
Long before this time you have found me out.

II.

VI. Ænigma 450, by Mr. K. T. Brunton.

I am frequently found
In a deal of good Ground ;
I travel all Day at my Leisure:
But at Night I lie still,
And of Reft take my fill,
Nor move but at fomebody's Plea-
If you take me in Hand, [fure.
Then I follow Command,
Lead me on where you pleafe, I don't
mind, Sir,
But at firft I'll not fail
For to make you turn Tail,
And then I'll lag on clofe behind, Sir.

A Phyfician I am
Without any Plan,
The Gravel I cure, an it pleafe you;
Without going to France,
Or learning to dance,   [eafy.
Your Walk I improve and make
What's rough and uncouth
I make pleafant and fmooth ;
I'm quiet if you let me alone, Sir :
If you afk for my Birth,
I fprung out of the Earth,
And my Mother was cut for the
Stone, Sir.

III. (1764) 1809 eng 912 ld p26

I'm a fingular creature, pray tell me my name,
I partake of an Englifhman's freedom and fame ;
I daily am old, and I daily am new,
I am prais'd, I am blam'd, I am falfe, I am true;
I'm the talk of the nation, while ftill in my prime,
But forgotten when once I've outlafted my time.
In the morning no Mifs is more courted than I,
In the evening no tuy thrown more carlefsly by.
Take warning, ye fair ! I like you, have my day,
And, alas ! you, like me, muft grow old and decay.

IV. (1780) Enigma 611

Tho' hard my fate I once knew hap-
pier times,   [climes ;
In fairer regions and in milder
Near where the fabled fauns and fa-
tyrs rove,   [and love ;
Where all was foftest pleafure, peace,
Vouchfaf'd my fervice to the ruftic
throng,   [z'd the fong ;
Their pleafure heighten'd, harmoni-
My friendfhip deign'd to all the
neighb'ring plain, [in my reign ;
And flocks and herds feem'd happy
My kindnefs felt when rolling tem-
pefts hurl'd   [world ;
Their bitt'reft fury o'er a weeping
Or when the fun diffus'd his warmer
ray,   [of day.
And pour'd on all the fervent flood
But now my woes commence, fad
fcene of grief !   [relief ;
And now behold me lab'ring paft
Behold my quiv'ring members beat
the ground,   [around.
And hear my dying murmurs all
Next chang'd my name, my fhape ro-
tund before,

Will now affume the pleafing form
no more.
Oblong my form, tho' curv'd my top
has been,   [feldom green.
Array'd in white, brown, blue, but
And tho' when living ne'er had fight
allow'd,   [the crowd.
Have fometimes eyes as 'twere to view
Nor ends my fhame tho' dead, in irons
hung,   [throng ;
Expos'd to th' infults of the giddy
And well remember at the midnight
heur,   [roar.
As Jack paft by me with a hideous
Scar.e could he hold direct his tot-
t'ring pace,   [face.
Tho' with foul dirt befpatter'd all my
Your favours, O ye fair, I fometimes
gain,   [years of pain ;
Which more than recompence whole
Ye deign/me oft the balm, the plea-
fing blifs ;   [to kifs :
If not your lips, your hands or robe
I fly to pleafe you, ftand to guard
your charms,   [my arms.
And oh ! how fafe you reft within

V. (1764)

Ladies, your moſt obedient——pray attend
To your Admirer, and devoted Friend.
Who courts your Favour—to your Service true,
Scarce knows a Pleaſure but what ſprings from you:
My Shape is ——— but no matter what ; my Mind,
Moſt Men agree, is of the Spaniel-kind ;
And, if you will truſt the Poets, I am blind,
But Poets, never in their Terms minute,
Heed not the Expreſſion, if their M rals ſuit ;
Beſide, their Judgment oft' gives place to Wit———
Hence, I am blind, cauſe B.unders I commit :
But to ſpeak freely, my Misfortune's ſuch,
Always to ſee too little, or too much———
Ladies, be kind, and pardon it as ſuch.
    To you I'm modeſt, undeſigning, free,
Kind, gen'rous, humble, or 'tis none of me ;
For know, a Sharper oft for me does paſs,
Cloſe, crafty, knaviſh——with a Front of Braſs ;
Ungen'rous, falſe — throughout an arrant Cheat,
Yet you, ye *Fair*, oft favour the Deceit ;
But ah ! beware ; for know, his only Aim
Is to deceive you, and blaſt your Fame.
    Ladies; ariſe, aſſert my injured Cauſe———
'Tis you muſt right me — you direct the Laws :
Drive the Uſurper from his ſoft Retreat,
The Bower of Bliſs, where earthly Joy's compleat.

VI. (1840)

Courage, ladies, don't be daunted,
Hear my story whilst I stay ;
Fly not, as with terrors haunted,
Your companion night or day.
I'm a brisk and lively fellow,
Tho' my gambols mayn't delight
    you ;
Frisky as a punchinello,
Let not my appearance fright you.
Yesterday and in my presence
Love imprinted burning kisses ;
Fondly call'd you bright quintes-
    sence,
Loveliest, fairest, best of misses.
I, inspir'd by Love's example,
Once a glowing kiss essay ;
Coyly you disdain the sample,
Shivering, frowning turn away.

Quick the frowner calls fair Dolly,
Instant bids her turn me out.
All her efforts are but folly,
I so nimbly skip about.
Now i'th' clutch of this cross vixen.
If I beg, and pity crave,
Rogue ! 'tis me you've play'd your
    tricks on ;   [save.
Nought on earth your life shall
Yet not in every place abused ;
Millions of us live together
Unmolested, ne'er ill used,
Fearing neither wind nor weather.
Then, fair ladies, why this coyness
Shewn to our devoted race ?
Here, appealing from your shy-
    ness,
Beg with Di a name and place.

VII. (1826) a Rebus

A chaplet meet for poet's head ;
A tree that's sacred to the dead ;
A sign of mercy,—England's
    pride,

Her guardian o'er the briny tide ;
The initials join, and you will see
A bard renown'd for poesy.

Appendix A:
1763 hut p224 225 ii 56
1764 e gima 160.hut v 233
1780 ld p29
1826 rebus p.21
1833 enigma 1155 p27
Answers:

  I. The letter "O"

 II. A Lawn roller

III. Morning newspaper

 IV. A door

  V. Time

 VI. A flea

VII. Byron [the poet]

Appendix B

Sample of a Complete Set of *Ladies' Diary* Mathematical Exercises

Questions published in No. 117 *of Ladies' Diary for* 1820.

1. What number is that which differs least from its common logarithm? and what number differs least from its hyperbolic logarithm?

2. Given the segments of the base made by the perpendicular from the vertical angle, to construct the plane triangle, when the *tangent* of one of the angles at the vertex made by the perpendicular has a given ratio to the *sine* of the other.

3. A cistern containing 5236 cubic inches, being filled with water, had dropped into it five heavy balls, whose diameters were in arithmetical progression. Required, the several diameters of the said balls; it being known that half the water was expelled in consequence of their immersion, and that the common difference of the terms of the progression was one inch.

4. Find two integers, such that their sum shall be a square, and their difference a cube number; but, if each of them be doubled, their sum shall be a cube, and difference a square number.

5. The three edges of a triangular pyramid which terminate in the vertex, are 12, 14, and 15; its perpendicular altitude 9; and the edges of its base are as the numbers 2, 4, and 5. Now, the distance of the centre of gravity from that angle of the base where the longest slant edge meets the two longest sides of the base being 6, what is the solidity of the pyramid?

6. Find the sides and areas, in whole numbers, of three scalene triangles, such, that their perimeters shall be equal, and their areas as the numbers 2, 7, and 15.

7. Give, by means of right lines and a circle, a general construction for the indeterminate equation $a^2 - 2a(x+y) + x^2 + y^2 + xy = 0$.

8. A person has to cross an elliptical common, the axes of which are eight and six miles, and has to call at a house situated in one

of the foci. On one side of the house he can walk at the rate of five miles an hour ; but, on the other, only four miles an hour. Required the least time in which he can perform his rectilinear journey across the common.

9. Demonstrate, synthetically, that an arc of any curve cannot be of finite curvature, unless the subtense ultimately vary as the square of the arc.

10. In order that the eclipses of Jupiter's satellites may be visible at any place, the planet must be at least $8°$ above, and the sun $8°$ below, the horizon. Now, on the 1st of February 1821, there will be an eclipse of Jupiter's third satellite, the emersion taking place at $6^h 19^m 3^s$ P.M. Greenwich time: can that emersion be observed at Berlin, N. lat. $52° 32' 30''$, E. lon. $13° 26' 15''$? Sun's declin. S. $17° 6'$. Jupiter's declin. S. $3° 31'$. Passage merid. $2^h 39^m$ P.M.

11. Suppose BC to be perpendicular to the horizontal line AC, and both of them to be given in length, and that a string of a given length (greater than $\sqrt{(AC^2 + BC^2)}$) is fastened at its two extremities to the points or tacks A, B ; and that the said string, as it is moved round A, is stretched tight by the continual motion of a point along the horizontal plane. Required, the curve that will thus be described on the horizontal plane, by the said stretching or tending point.

12. What will be the ratio of the forces of gravity, at the surface of the earth, at the top of a slender cylinder a mile high, at the upper surface of a sphere a mile in diameter (in contact with the earth and of the same medium density), and on the top of an extensive piece of table-land of the same density and a mile high, taking the earth's radius at 3960 miles?

13. In the great solar eclipse which happens Sept. 7th, 1820, the apparent time of the greatest obscuration at Greenwich is $1^h 52^m 48^s \cdot 2$ P.M. when the angular distance about the centre of the sun from its vertex to the centre of the moon is $17° 18' 22''$ to the left hand ; and the visible distance of their centres to the difference of their horizontal parallaxes as $0 \cdot 5665946$ to 1. Hence it is required to find the direction and distance from the above place, which a person must travel for the shortest *journey* to observe the sun centrally eclipsed ; and at the same time the angle of elevation, and the direction he must ascend in a balloon for the shortest *voyage* to be gratified with a sight of the same phænomenon ; supposing the earth a perfect sphere, and its diameter 7914 miles?

14. It is required to investigate a theorem, comprehending the pressure on the base of a steam-engine piston, the course or stroke of that piston and its velocity, on one part; and the velocity,
mass,

mass, and diameter of a fly, on the other part; so that one rotation of the fly, with its initial velocity, shall produce a dynamic effect equal to that of the piston in n successive strokes.

N. B. The mass of the fly is supposed equally distributed over its rim, and the diameter of the crank handle equal to the course of the piston.

15. The major and minor axes of an elliptical billiard-table are $2a$ and $2b$. Suppose an elastic ball to be propelled through one of the foci perpendicularly to the major axis, what will be the rectangular co-ordinates which indicate its position at the tenth reflection? and will it, after any finite number of reflections, move to and fro in the direction of the major axis?

Answers:

1. 0.4542944819 and 1.

2.

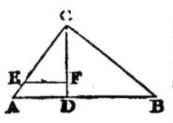

Let ABC be the triangle required, CD the perpendicular. Take CE = CD, and draw EF perp. to CD: then DB is the tangent of the angle DCB to radius CD, and EF is the sine of ACD to the same radius. Per question the ratio of DB to EF is given, and DB is given; therefore EF is given. So that we have given the sine EF, and the tangent AD of an angle ACD, to find the angle, which has been done in the solutions to Qu. 1. L. Di. 1819. This being determined, the construction of the triangle is obvious.

3. 7.80087, 8.80087, 9.80087, 10.80087, 11.80087.

4. 132.124.

5. 21.48785.

6. Perimeter: 1092, Areas: 5460, 9110, 40590.

7. See construction LD.

8. 1.00936 hr.

9. See discussion LD.

10. Distance from zenith is 700 57 min. 47 sec., altitude 100 2 min. 13 sec. Eclipse is visible.

11.

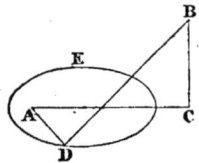

Suppose A, B, to be the places of the two tacks, and imagine any plane whatever to pass through those two points; then, if the string is stretched tightly, and a point carried along it, on that plane, it will, as is well known, describe the curve of an ellipse; and so of every other plane. That is, if this ellipse be conceived to revolve upon the diameter which passes through A, B, it will generate a spheroid, in some point of the surface of which the describing point D will be found in every conceivable position of the stretched thread. Let this spheroid be cut by the horizontal plane which passes through A, C, perpendicular to B, C, then (*Hutton's Mensuration*, prop. 1. sect. ii. part iii.) the section is *an ellipse;* which is, consequently, the curve required.

Had BC coincided with BA, the section of the spheroid would have been perpendicular to the axis, and the curve described *a circle.*

Otherwise, by Messrs. Anthony Cook, T. Hudspeth, W. Oakes,

12. Let f = force of gravity at surface of Earth, then answer: 0.9998741f.

13. The computation for this problem was so complex and lengthy that the editor of the *Diary* commented on this situation. While it was not published, the answer was provided. This problem was correctly solved by nine correspondences. The correct answer they arrived at is given below:

> The place nearest to Greenwich (on the earth's surface) where the sun was centrally eclipsed, Sept. 7, 1820, is in lat. 54° 53′ 37″·7 north, long. 6° 51′ 41″·8 east. To arrive at this place, a person must travel N.E. 2° 34′ 16″·9 eastward, a distance of 368·8705529 miles.
>
> The length of the shortest aërial *voyage*, to effect the same purpose, is 224·2014832 miles, the angle of ascent above the horizon of Greenwich 48° 0′ 47″·9, and the direction N.E. by E. 7° 17′ 3″·1 east.

14. See discussion LD.

15. See discussion LD.

Consult *Ladies' Diary*, 1821, pp.33-47 for solution techniques and extended discussions as required.

Appendix C

Selected Problems with Worked Solutions

For educational purposes, one of the most outstanding features of *The Ladies' Diary* and its resulting compilations was that they supplied mathematical feedback on the solutions. They demonstrated correct worked-out answers. The following examples have been chosen from Leybourn's collection, 1817 (Ley) or actual issues of the *Diary* (LD). The extensive discussion on Newton's problem is included in its entirety to demonstrate the scope of interactions. The excerpt of Mr. Skene taken from The *Gentleman's Mathematical Companion*, William Davis , editor, 1805, 2: No. 8: 8-18 [**Dav05**], is actually signed β. Cygni (a pseudoynom?).

Questions proposed in 1711, *and answered in* 1712.

I. QUESTION 17.

I happen'd one ev'ning with a tinker to sit,
Whose tongue ran a great deal too fast for his wit:
He talk'd of his art with abundance of mettle,
So I ask'd him to make me a flat-bottom kettle,
That the top and the bottom diameters be
In just such proportion as five is to three.
Twelve inches the depth I would have and no more,
And to hold in ale gallons seven less than a score.
He promis'd to do't, and to work he strait went;
But when he had done it, he found it too scant.
He alter'd it then, and too big he had made it,
And when it held right, the diameters fail'd it:
So that making't so often, too big, or too little,
The tinker at last had quite spoil'd the kettle:
Yet he vows he will bring his said purpose to pass,
Or he'll utterly spoil ev'ry ounce of his brass.
To prevent him from ruin, I pray help him out,
The diameter's length else he'll never find out.

Answered by Mr. Richard Parker.

Lesser diameter 14·6390238 ; greater diameter 24·398373 inches.

First Solution.

Putting $5z$ and $3z$ for the top and bottom diameters, by page 525 Mensuration, we shall have $(3 \times 64zz + 4zz$ or) $196zz \times 12 \times \cdot00023209 = 13$: hence $z = \sqrt{\dfrac{13}{196 \times 12 \times \cdot00023209}}$

$= 4\cdot880057$; consequently $3z = 14\cdot640171$, and $5z = 24\cdot400285$, the two diameters required. H.

Second Solution.

Find the content of the frustum of a cone, whose altitude is the same as that of the given one, and the diameters of the top and bottom 5 and 3. This content will (by prob. 6. vol. ii. page 48. of the course) be $= \frac{1}{3} \times (5^2 + 3^2 + 5 \times 3) \times \cdot7854 \times 12 = 153\cdot982$ cubic inches. The content of the given frustum is $= 13 \times 282 = 3666$ cubic inches. Because the frustums have the same altitude, they are to one another in the same ratio as the areas of their ends, or as the squares of the diameters of their ends. Hence $153\cdot982 : 3666 :: 5^2 : 595\cdot3687$, the square of the diameter of the greater end ; the square root of which is $24\cdot4002$, the greater diameter : and as $5 : 3 :: 24\cdot4002$ the greater diameter : $14\cdot6401$, the less diameter. L.

Third Solution.

Let the given frustum be completed into a cone. Then the part added will be a cone similar to the whole cone. Let D and d denote the diameters of the greater and less ends of the given frustum respectively ; P and p the perpendiculars of the whole cone and the part added, the contents of these cones being denoted by c and c. Then by the question $5 : 3 :: D : d$, but by similar triangles $D : d :: P : p$; therefore $5 : 3 :: P : p$, and therefore $5 - 3 : 3 :: P - p$ (or 12 the height of the frustum) : p, the height of the part added $= 18$; hence P $= 18 + 12 = 30$ inches, the height of the whole cone. Because similar solids are to one another in the same ratio as the cubes of their altitudes, we have as $P^3 : p^3 :: c : c$, therefore $P^3 - p^3 : P^3 :: c - c$, (the given frustum) : c ; that is, as $27000 - 5832$ or $21168 : 27000 :: 3666 : \dfrac{3666 \times 27000}{21168} = c = 4676\cdot02$, the content of the whole cone. Hence, because $\cdot7854 \times D^2 \times \frac{1}{3}$ P $= c$, $D = \sqrt{\dfrac{c}{\frac{1}{3} \times \cdot7854 \times P}} = \sqrt{\dfrac{4676\cdot02}{\frac{1}{3} \times \cdot7854 \times 30}} = 24\cdot4002$, the greater diameter. And $5 : 3 :: 24\cdot4002 : 14\cdot6401$, the less diameter. L.

The ingenious Mrs. Babara Sidway, in her answer to this question, proposed another very pretty

QUESTION.

If the frustum of the cone was to hold as much again : what would be the length of the part added to the greater end ?

First Solution to Mrs. Sidway's Question.

If $5a$ and $3a$ be the diameters of the first or given frustum, whose altitude is 12, and z the altitude of the frustum to be added : Then, by similar triangles. &c. $\frac{1}{6}az + 5a =$ the greater diameter of the part

QUESTION.

If the frustum of the cone was to hold as much again : what would be the length of the part added to the greater end ?

First Solution to Mrs. Sidway's Question.

If $5a$ and $3a$ be the diameters of the first or given frustum, whose altitude is 12, and z the altitude of the frustum to be added : Then, by similar triangles. &c. $\frac{1}{6}az + 5a =$ the greater diameter of the part

added; and since the two contents are equal, we shall have $((\frac{1}{6}az + 5a)^2 + (\frac{1}{6}az + 5a) \times 5a + 25a^2) \times z = (25aa + 15aa + 9aa) \times 12$, or $((\frac{1}{6}z + 5)^2 + (\frac{1}{6}z + 5) \times 5 + 25) \times z = 49 \times 12 = 588$, or $z^3 + 90 z^2 + 2700z = 21168$. Hence $z = 6.3847619 =$ the height of the part to be added. H.

(Ley, 1817, vol. I: pp.13-15.)

Second Solution to Mrs. Sidway's Question.

Let the given frustum be completed. Let the height and content of the whole cone be found as in the last solution to the original question : The height is 30 inches, and the content of the whole cone is 4676·02 cubic inches. To this add 3666 cubic inches, the content of the given frustum; the sum is 8342·02 cubic inches, which is the content of the cone, including the lengthened frustum. Then since similar solids are to one another as the cubes of their altitudes,

we have as $4676 \cdot 02 : 8342 \cdot 02 :: 30^3 : \dfrac{8342 \cdot 02}{4676 \cdot 02} \times 30^3$, the cube of

the altitude of the last mentioned cone ; the cube root is $30 \times$

$\sqrt[3]{\dfrac{8342 \cdot 02}{4676 \cdot 02}} = 36 \cdot 382$ nearly, from which taking 30, (the altitude

of the cone, including the given frustum), there remains 6·382 for the length of the frustum to be added to the greater end of the given frustum, so that it may hold twice the quantity as at first. L.

One of the most interesting, in the sense of the solutions offered and the resulting interactions, is the Prize Mathematical Question confronted in the 1744 issue of the *Diary*:

THE PRIZE QUESTION, *by the late illustrious* Sir I. Newton.

Three staves being erected, or set up on end, in some certain place of the earth, perpendicular to the plane of the horizon, in the points A, B, and c; whereof that which is at A, is 6 feet long; that in B, 18; that in c, 8; the line AB being 33 feet long : It happens on a certain day in the year, that the end of the shadow of the staff A passes through the points B and c: and of the staff B, through A and c; and of the staff c, through the point A.

To find the sun's declination, and the elevation of the pole or day, and the place where this shall happen.

Note, this is the 42d problem in Sir Isaac Newton's Universal Arithmetic; and it may seem a piece of vanity in attempting to give a solution after the greatest of men; but having in the winter 1740, taken a great deal of pains to bring out a solution, and never being able to get his numbers for the declination and latitude precisely the same, I was fond to think his were exact, and wrought it over and over again; at first it came out an adfected equation, then a quadratic, and at last happily by a simple equation; and having taking the pains to prove all the numbers (not depending on the logarithms) found them agree in every particular, and by construction to form a true conic section. We therefore humbly presume, that in a calculus so prolix and difficult (in Sir Isaac's method) there might happen a small error, or at least some press fault of the editions, or in the translation; which we hope to make more fully appear in the next year's Diary.

Answered by Mr. Ant. Thacker.

Put $a = 6$, $b = 18$, $c = 8$ feet, the height of the three staves at A, B, c; the line AB $= 33$ feet $= d$; the distance between the top of the staff A and bottom of the staff B ($= aB$) call e; the distance between the top of the staff B and bottom of staff A ($=bA$) call f; then will $\dfrac{(c + a)\,(e + f)}{2b + 2a} + \dfrac{(b + a)\,(c - a)}{2c + 2f} = m = 21 \cdot 0789$, the distance from the top of c to the bottom of A. Then (by 47 Euc. I.) AC is found $= 19 \cdot 501$, which call x; also $\dfrac{(c + b)\,(e + f)}{2b + 2a} + \dfrac{(b - c)\,(b + a)}{2e + 2f} = n = 40 \cdot 216$, the length between the top of B and bottom of c; and hence CB is found $= 35 \cdot 962 = z$. Now putting $\dfrac{be - fa}{df + de} = h = 16111$; $\sqrt{(xx + aa)} = s = 20 \cdot 404$; $\dfrac{cs - am}{xm + xs} = k = \cdot 0454$; and $\dfrac{dd + xv - zz}{2dx} = v = \cdot 13501$; then will

$$\frac{2-2vv}{1+hh+kk-2hkv-vv} = 1 \cdot 948315, \text{ the versed sine of an arc}$$

which is double the latitude 161° 30′; whose half is $= 80^\circ\ 45'$, **the** true latitude required.

Now calling the sine of the latitude p, then will $\dfrac{pc + pa}{s + m} =$
·33309 $=$ the sine of 19° 27′, the sun's declination **north.**

Sir I. Newton, the author of this problem finds the lat. 80° 45′ 20″, and declination 19° 27′ 20″, as may be seen in the 42d prob. of his Universal Arithmetic, in the English edit. 1720, p. 151, where the translator Mr. Raphson, Mr. Cunn, or the printer, committed a blunder, making the line AB $=$ 30, instead of 33. By reason of which disappointment, the solution is here shorter than was designed; but the investigation of the theorem above we have printed in the first volume of Diary Questions [Thacker's Miscel.] If the line AB was $=$ 30, then the latitude is 80° 4′ 7″, and declination 21° 7′ 4″

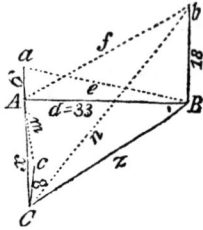

Here follows the Solution from Thacker's *Miscellany alluded to above.*

Call height of the staff in the point A, (a); that in B, (b); that in C, (c); the line AB (d); the distance betwixt the top of the staff in A, and the bottom of the staff in B, (e); the distance betwixt the top of the staff in B, and the bottom of the staff in A, (f); and let AC $= x$, CB $= z$; then $\sqrt{(xx + aa)} = s$, and $\sqrt{(xx + cc)} = m$, the distances betwixt the top of the

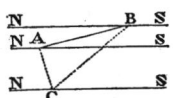

staff in A and bottom of the staff in C, and top of the staff in C and bottom of the staff in A, respectively. And suppose NS the meridian line, radius $= 1$; then will $a \div e$ and $d \div e$ be the sine and cosine of the sun's altitude when he makes the shade AB by the staff A; for the cosine of the angle ABN (or BAN) $=$ cosine of the sun's azimuth from the north at that time, put $\mp v$. Again, $b \div f$ and $d \div f$ express the sine and cosine of the sun's altitude when he makes the shade BA by the staff B; the cosine of BAN (or ABN) $=$ cosine of the sun's azimuth from the north at that time $\mp v$. Also $a \div s$ and $x \div s$ will be the sine and cosine of the sun's altitude when he makes the shade AC by the staff A; and let $\mp y$ be put to denote the cosine of the angle ACN (or CAN $=$) the sun's azimuth from the north at that time; again, $c \div m$ and $x \div m$ will express the sine and cosine of the sun's altitude when he makes the shade CA by the staff C; and $\mp y$ the cosine of the angle CAN (or ACN) $=$ the cosine of the sun's azimuth at that time from the north. Lastly, let p and q be put to express the sine and cosine of the required latitude; and g, the sine of the sun's declination. Then (1) $\dfrac{pa}{e} \mp$

$\dfrac{qdv}{e} = g$; (2) $\dfrac{pb}{f} \mp \dfrac{qdv}{f} = g$; (3) $\dfrac{pa}{s} \mp \dfrac{qxy}{s} = g$; (4) $\dfrac{pc}{m} -$

$\dfrac{qxy}{m} = g$. Whence, by equation (1), we have $\pm qdv = ge - pa$; and by equation (2) $pb - gf = + qdv$; consequently $ge - pa = pb -gf$; and $g = \dfrac{pb + pa}{e + f}$; again, by equation (3), we get $\mp qxy = gs - pa$, and by equation (4), $pc - gm = + qxy$: consequently $gs - pa = pc - gm$, $\therefore g = \dfrac{pc + pa}{s + m}$; therefore

$\dfrac{pc + pa}{s + m} = \dfrac{pb + pa}{e + f}$, and $\dfrac{c + a}{s + m} = \dfrac{b + a}{e + f}$; or $\dfrac{(c + a)\,(e + f)}{b + a}$

$= s + m$; $\therefore \dfrac{(c + a)\,(e + f)}{b + a} - m = s$, which squared is

$$\frac{(c + a)^2 (c + f)^2}{(b + a)^2} - \frac{2m(c + a) (c + f)}{b + a} + m^2 = s^2:$$ but as

$\sqrt{(x^2 + a^2)} = s$, and $\sqrt{(x^2 + c^2)} = m$, $\therefore (x^2 =) s^2 - a^2 = m^2 - c^2$, $\therefore s^2 = m^2 + a^2 - c^2$, which put for s^2 in the above equation,

makes $\dfrac{(c + a)^2 (c + f)^2}{(b + a)^2} - \dfrac{2m(c + a) (c + f)}{b + a} + m^2 = m^2 + a^2$

$- c^2$, $\therefore$ by transposition, &c. we have $\dfrac{(c + a) (c + f)}{2b + 2a} +$

$\dfrac{(b + a) (c - a)}{2c + 2f} = m$ the distance betwixt the top of the staff in c

and bottom of the staff in A, and so AC is found.

And by the very same way that m was found, the distance betwixt the top of the staff in B and bottom of the staff in c may be found, and if n be put for it, the equation will stand $\dfrac{(c + b) (c + f)}{2b + 2a} +$

$\dfrac{(b - c) (b + a)}{2c + 2f} = n$, and therefore CB is known. For the other part of the analysis, we have, by the help of equations (1) and (2) $\dfrac{pa}{e} \pm \dfrac{qdv}{e} = \dfrac{pb}{f} \mp \dfrac{qdv}{f}$ $\therefore \pm \dfrac{qdv}{e} \pm \dfrac{qdv}{f} = \dfrac{pb}{f} - \dfrac{pa}{e}$ $\therefore$

$\pm v = \dfrac{(be - fa)p}{(df + dc)q}$; put $\dfrac{bc - fa}{df + de} = h$, then will $\pm v = \dfrac{hp}{q}$, whose

square is $v^2 = \dfrac{h^2 p^2}{q^2}$ $\therefore \sqrt{(1 - v^2)} = \sqrt{\left(1 - \dfrac{h^2 p^2}{q^2}\right)}$ = the sine of

the angle BAS. And by the same way of reasoning, from equations (3) and (4), is found $\pm y = \dfrac{(cs - am)p}{(xm + xs)q}$; put $\dfrac{cs - am}{xm + xs} = k$, then

will $\pm y = \dfrac{kp}{q}$, which squared makes $y^2 = \dfrac{k^2 p^2}{q^2}$ $\therefore \sqrt{(1 - y^2)} =$

$\sqrt{\left(1 - \dfrac{k^2 p^2}{q^2}\right)}$, the sine of the angle CAS.

Now having all the three sides of the triangle ABC given, we find the angle CAB to be less than a right angle; therefore the cosine belonging to the sum of the angles BAS and CAS is $\dfrac{hkp^2}{q^2} - \sqrt{(1 - }$

$\dfrac{h^2 p^2}{q^2})\sqrt{\left(1 - \dfrac{k^2 p^2}{q^2}\right)}$. But from the triangle we have $\dfrac{x^2 + d^2 - z^2}{2dx} =$

v the cosine of the angle CAB, consequently $\dfrac{hkp^2}{q^2} - \sqrt{\left(1 - \dfrac{k^2 p^2}{q^2}\right)}$

$\sqrt{\left(1 - \dfrac{k^2 p^2}{q^2}\right)} = v$ $\therefore \dfrac{hkp^2}{q^2} - v = \sqrt{\left(1 - \dfrac{h^2 p^2}{q^2}\right)}\sqrt{\left(1 - \dfrac{k^2 p^2}{q^2}\right)}$ which squared, &c. makes $- \frac{2khvp^2}{q^2} + v^2 = 1 - \frac{h^2p^2}{q^2} - \frac{k^2p^2}{q^2}$, or, by

writing $1 - p^2$ for q^2, $- \frac{2hkvp^2}{1-p^2} + v^2 = 1 - \frac{h^2p^2}{1-p^2} - \frac{k^2p^2}{1-p^2}$.

And by reduction $2p^2 = \frac{2 - v^2}{1 + h^2 + k^2 - 2hkv - v^2}$ the versed sine

of an arc, which is double the required latitude. And the sine of the

sun's declination is known to be $g = \frac{n\sigma + pa}{s + m}$.

Another Solution, by Mr. Skene, taken from Davis's Mathematical Companion for 1805.

Among a variety of problems, which, were proposed, about the time of Des Cartes, as instances of the excellence of the modern analysis, and its superiority to the ancient, the following appears to be the most abstruse, as well as the most curious and remarkable. Des Cartes * speaks of it in his Epistles as a problem of the greatest difficulty, and eminently fitted ad notandam industriam bene disscrendi equationes. From Francis à Schooten we learn that it was first published in the year 1640, in a very ingenious book, entitled Den Onwissen Wis-konstenaer I. I. Stampioënius. † The solution there exhibited, is revised, corrected, and improved in Schooten's Additamentum to his Commentary on Des Cartes's Geometry. ‡ The Additamentum, indeed, is principally occupied with this problem, which is placed at the end of the Commentary as a proof, as he himself expresses it, non facile problema aliquod datum iri, quod hanc geometriam affugiat, aut ejusdam methodo solvi non possit. Sir Isaac Newton has likewise given a solution to this problem in his Arithmetica Universalis, published by Whiston 1707. It may be justly concluded, that it would be in vain to seek for a more elegant solution of a problem which has passed through the hands of such eminent men. Newton's, in particular, is one of the finest specimens of algebraic analysis in the whole circle of the mathematical sciences. But as the question has been supposed to be beyond the reach of the ancient geometry, I shall make no apology for giving the following solution, which is effected without the assistance of algebra.——The problem is thus enunciated by Des Cartes.—— "Tres baculi erecti sunt ad perpendiculum in horizontali plano, ex punctis A, B, c; et baculus A est 6 pedum, B 18 pedum, c 8 pedum, et linea AB est 33 pedum; et una atque eadem die extremitas umbræ solaris, quam facit baculus A, transit per puncta B & c, extremitas umbræ baculi B per A & c, et ex consequenti etiam baculi c per A & B. Quæritur in quænam poli altitudine, & qua die anni it

_____

* Ren Des Cartes, Epist. Part II. Epist. LXXI. page 257, Francf. Ed.
† Edited by Jacob à Waessenaer. L.
‡ Schooten merely premised a lemma and added an explanation by Erasmius Bartholinus. L.

contingat." This enunciation is in part erroneous, for it does not
follow, as is here affirmed, that the end of the staff at c must pass
through the points A and B, because the end of the staff at A passes
through B and c, and that of the staff at B through A and c: another circumstance must be given, namely, that the end of the staff
at c passes through A, so that it is necessary to change " et ex consequenti, &c. into baculi c per punctum A, et ex consequenti etiam
baculi c per punctum B." In this manner it is enunciated by Schooten, with this difference only, that the consequence from the other
part of the data the end of the staff at c's passing through B is demonstrated by him algebraically ; and the same is done by Newton.

The following is a geometrical demonstration.

The end of the shadow of the staff at A describes a conic section,
which let be CEDGB (fig. 1.). Let A, B, c, be the
places of the three staves, and let BA, CA be produced
to meet the curve in D, G. Draw also EF through
A parallel to BC, to meet the curve in E, F, and
join DE, GF. Then when A casts the shadow AD,
B casts the shadow BA ; and when A casts AE, B
casts BC. Therefore AD is to BA as AE to BC ;
but AE is parallel to BC by the construction, consequently DE is parallel to AC. Now BC being parallel to EF, CG parallel to ED, and c. E two points
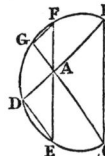
in the conic section, GF is parallel to BD, by Simson's Con. Sect.
lib. iv. prop. 29. Hence BC is to AF as CA to AG, and the end of the
shadow of c therefore passes through the point B, whatever be the
conic section described by the extremities of the shadows. *Q. E. D.*

This property in the ellipse may be still more easily demonstrated
by first shewing its truth in the circle, and thence transferring it to
the ellipse by cutting a cylinder obliquely.

We shall now proceed to point out our method of resolving the
problem, for which purpose the following lemma is required.

Lemma. If the altitudes of the sun be taken at the same place on
the same day, when he is in two opposite directions, the sum of
their tangents will be to the sum of their secants as the sine of the
sun's declination to the sine of the latitude of the place.

Let z (*fig.* 2.) be the zenith, p the pole, zp the co-latitude, zq, zo the co-altitudes, and pq or po the comple-ment of the sun's declination. Draw pm per-pendicular to zo, and qm, om will be equal to each other; therefore om is equal $\frac{1}{2}$(zo + zq) and zm equal to $\frac{1}{2}$(zo — zq). Then, by an elementary property of spherical tri-angles, cos pq : cos zp :: cos om : cos zm :: cos $\frac{1}{2}$(zo + zq) : cos $\frac{1}{2}$(zo — zq). But cos $\frac{1}{2}$(zo + zq) is to the cos $\frac{1}{2}$(zo — zq), as the cot zo + cot zq to the cosec zo + cosec zq; therefore cos pq : cos zp :: cot zo + cot zq : cosec zo + cosec zq, that is, the sine of the sun's decli-

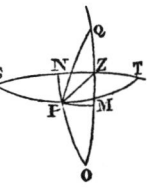

nation is to the sine of the latitude of the place as the sum, &c. *Q. E. D.*

Cor. If two altitudes of the sun be taken on the same day, when he is in opposite directions, and other two altitudes when he is like-wise in opposite directions, the sum of the tangents of the first two will be to the sum of the tangents of the second two, as the sum of the secants of the first two to the sum of the secants of the second two. This corollary follows directly from the lemma, as the latitude and declination remain the same.

Now as the heights of the two staves at a and b (*fig* 3.) are given, and the lengths of their respective shadows, when the sun is in opposite directions, the two altitudes are likewise given, and consequently the ratio of the sum of their tangents to the sum of their secants. Let ad represent the staff at a, and ce the staff at c. Take cf on ec produced

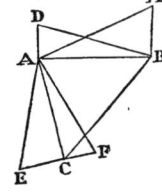

equal to ad, and join ae, af. Then will ce and cf represent the tangents, and ae, af the secants of the sun's altitudes, when in the op-posite directions ac, ca; and therefore by the lemma the ratio ef to ae + af is given; but ef is given, consequently ae + af is given.

Now, by an elementary proposition in plain trigonometry, ae + af is to ef as the difference between ce, cf to the difference between ae, af; which difference is therefore given, and consequently ae, af. Hence ac, the distance between the two staves a and c, becomes known.

Then having given the three sides of the triangle abc, the angle bac may be determined; and as the heights of the staves at a and c are given, and their distance ac, the sun's altitudes, when in the op-posite directions ca, ac, are likewise given.

Now let z on the globe represent the zenith of the place, (*fig.* 2.) at which the observation is made, p the pole, zp the co-latitude, zq, zo the zenith distances of the sun, when in the opposite directions ab, ba, and zs, zt the zenith distances of the sun, when in the op-posite directions ca, ac. Draw the great circles pq, po, ps, pt, which will be all equal, being the co-declination of the sun: also pm perpendicular to oq, and pn to st.

It is evident that zm, which is equal to $\frac{1}{2}$(zo — zq), and zn, which is equal to $\frac{1}{2}$(zs — zt), are both given, but nzm is equal to the angle bac (*fig.* 3.), being the difference of the sun's azimuths when in the directions ab, ac. Hence this construction.

Take zm in zo equal to $\frac{1}{2}$(zo — zq), and zn in zs, whose position is given, equal to $\frac{1}{2}$(zs — zt); draw the great circle mp perpendi-cular to oq, and np perpendicular to st; the point p where they inter-sect each other is the pole, which being determined the latitude and declination are given.

We shall now give the numerical calculation.

Let AD *(fig. 3.)* perpendicular to AB be equal to the staff at A, BH perpendicular to AB equal to the staff at B, and let BD, AH be drawn. Then will AD $= 6$, BH $= 18$, CE $= 8$, and AB $= 33$. Hence the tan ABD $= \dfrac{2}{11}$, and sec ABD $= \dfrac{5\sqrt{5}}{11}$, also tan BAH $= \dfrac{6}{11}$ and sec DAH $= \dfrac{\sqrt{157}}{11}$. Therefore $\dfrac{2}{11} + \dfrac{6}{11} : \dfrac{\sqrt{125}}{11} + \dfrac{\sqrt{157}}{11} :: 8 :$ $\sqrt{125} + \sqrt{157} :: 1 : 2 \cdot 963788 ::$ sum of the tangents of the sun's altitudes : sum of their secants, when in any two opposite directions. Hence, EF being $= 14$, we get AE $= 21 \cdot 0839221$, AF $= 20 \cdot 4091099$, and AC $= 19 \cdot 5072236$. In like manner, we find BC $= 35 \cdot 963158$. Now in the triangle ABC the three sides are given, from which the angle BAC is easily computed $= 82^\circ\ 8'\ 5''$. We have also given *(fig. 2.)*, ZO $= 79^\circ\ 41'\ 43''$, ZQ $= 61^\circ\ 23'\ 22''$, ZS $= 72^\circ\ 54'\ 11''$, ZT $= 67^\circ\ 42'\ 1''$, and the angle SZO $=$ MZN $= 82^\circ\ 8'\ 5''$; hence ZM $= 9^\circ\ 9'\ 10\frac{1}{2}''$, ZN $= 2^\circ\ 36'\ 5''$, and by spherics cos PZM : cos PZN :: tan ZM : tan ZN : but the sum of the angles PZM, PZN is given $= 82^\circ\ 8'\ 5''$, and the ratio of their cosines, from which we easily find PZM $= 8^\circ\ 20'\ 8''$, and PZN $= 73^\circ\ 47'\ 57''$. Whence the latitude and declination may be computed by the following analogies, tan ZM : rad :: cos PZM : cotan ZP $=$ tan lat. and by the lemma $1 : 0 \cdot 3374061 ::$ sin lat. : sine declination. In this manner, we find that the latitude of the place was $80^\circ\ 45'\ 4''$, and the sun's declination $19^\circ\ 27'\ 9''$, which numbers agree exactly with those obtained in so different a manner by Francis Schooten and Sir Isaac Newton.

(Ley, 1817, I:pp.343-350)

Bibliography

[AB09]   Joe Albree and Scott H. Brown, *"A valuable monument of mathematical genius": The Ladies' Diary (1704–1840)* (English, with English and French summaries), Historia Math. **36** (2009), no. 1, 10–47, DOI 10.1016/j.hm.2008.09.005. MR2272883

[ACN57] Francesco Algarotti, Elizabeth Carter, and Isaac Newton, *The english common reader: A social history of the mass reading public, 1800–1900*, second ed., Columbus: Ohio State University Press, 1957.

[Adb72]   Alison Adburgham, *Women in print: Writing women and women's magazines from the restoration*, Allen and Unwin, 1972.

[Agn48]   Maria Agnesi, *Instituzioni analetiche [analytical institutions]*, 1748.

[AH08]   Amy Ackerberg-Hastings, *John Playfair on British decline in mathematics*, BSHM Bull. **23** (2008), no. 2, 81–95, DOI 10.1080/17498430802037830. MR2431845

[Alg39]   Francesco Algarotti, *Sir Isaac Newton's philosophy explain'd for the use of the ladies*, E. Cave, 1739, Elizabeth Carters's translation of Il Newtonianismo per le dame, 1737.

[Ami95]   Amir Alexander, *The imperialist space of Elizabethan mathematics*, Stud. Hist. Philos. Sci. **26** (1995), no. 4, 559–591, DOI 10.1016/0039-3681(95)00023-2. MR1368392

[Arb01]   John Arbuthnot, *An essay on the usefulness of mathematical learning*, 1701.

[Arc29]   Raymond Archibald, *Notes on some minor English mathematical serials*, Mathematical Gazette **14** (1929), 379–400.

[Asc01]   Roger Ascham, *The scholemaster*, John Daye, 1701.

[Ast94]   Mary Astell, *A serious proposal to the ladies*, T.W, 1694.

["B22]   "Bibliapola", *Female correspondents*, Leeds Correspondent **IV** (1822), 116–121.

[Bab30]   Charles Babbage, *Reflections on the decline of science in England*, B. Fellows, 1830.

[Bai17]   Melissa Bailes, *Questioning nature: British women's scientific writing and literary originality, 1750–1830*, University of Virginia Press, 2017.

[Bal89]   James Gow, *A short history of Greek mathematics*, Cambridge Library Collection, Cambridge University Press, Cambridge, 2010. Reprint of the 1884 original; Previously published by Chelsea Publishing Co., New York, 1968 [MR0238652]. MR2859188

[Bar34]   Isaac Barrow, *The usefulness of mathematical learning, explained and demonstrated, being mathematical lectures read in the publick schools at the University of Cambridge*, Reprinting, Frank Cass & Co. Ltd., London, 1970. MR0270887

[Bat87]   Robert Bataille, *Robert Heath, Thomas Cowper and eighteenth-century mathematical journalism*, Papers of the Bibliographical Society of America **1** (1987), 339–343.

[Bat90]   _____ , *Elizabeth Beighton and The Ladies' Diary*, Journal of Newspapers and Periodical History **6** (1990), 20–24.

[BC97]   Hannah Baker and Elaine Charles, *Gender in eighteenth-century England*, Longmans, 1997.

[Ber34]   George Berkeley, *The analyst: A discourse addressed to an infidel mathematician*, J. Tonson, 1734.

[BH79]   M. P. Black and A. G. Howson, *"A source of much rational entertainment"*, Math. Gaz. **63** (1979), no. 424, 90–98, DOI 10.2307/3616014. MR535859

[Bla61]   Cyprian Blagden, *Thomas Carnan and the almanack monopoly*, Papers of the Bibliographical Society of America **14** (1961), 23–43.

[Bla77]   _____, *The stationers company: A history, 1403–1959*, Stanford University Press (1977).

[Bon82]   John Bonnycastle, *Introduction to algebra*, 1782.

[Bow01]   Richard Bowden, *"The English stock and the stationers company: The final years."* The stationers' company: A history of the later years, 1800–2000, Worshipful Company of Stationers and Newspaper Makers, 2001.

[Bra59]   George Brauer, *The education of a gentleman, theories of gentlemanly education in England 1660–1775*, 1959.

[Bra95]   B Braithwaite, *Women's magazines: The first 300 years*, 1995.

[Bru76]   M.T. Bruck, *Riddling: Occasion to act*, Journal of American Folklore **89** (1976), 139–165, 352.

[Bru96]   _____, *Mary Somerville, mathematician and astronomer of underused talents*, Journal of the British Astronomical Association **106** (1996), 201–206.

[Bru15]   Melody Bruce, *Analyzing Melatiah Nash and the Ladies' and Gentlemen's Diary*, International Journal of Undergraduate Research and Creative Activities **7** (2015), 1–7.

[Caj19]   Florian Cajori, *A history of the concepts of limits and fluxions in Great Britain*, 1919.

[Cap79]   Bernard Capp, *Astrology and the popular press: English almanacs, 1500–1800*, Faber and Faber, 1979.

[Cas80]   J. W. S. Cassels, *The Spitalfields Mathematical Society*, Bull. London Math. Soc. **11** (1979), no. 3, 241–258, DOI 10.1112/blms/11.3.241. MR554388

[CB77]   Henry Clarke and J. E. Baily, *The school candidates: A prosaic burlesque*, 1877.

[Cer02]   Charles Cerame, *Benjamin Banneker: Surveyor, astronomer, publisher, patriot*, 2002.

[Cha73]   Hester Chapone, *Letters on the improvement of the mind addressed to a young lady*, 1773.

[Cha12]   Alexander Chalmers, *Chalmers biographical dictionary*, 1812.

[Cla29]   Frances Marguerite Clarke, *Thomas Simpson and his times*, ProQuest LLC, Ann Arbor, MI, 1929. Ph.D. Thesis, Columbia University. MR2936217

[Coc44]   Edward Cocker, *Arithmetick*, 1644.

[Col44]   John (translator) Colton, *Analytical institutions*, 1644, Translation of Maria Agnesi's Instituzioni Analetiche [Analytical Institutions], 1748.

[Coo99]   Alan Cook, *Henry Beighton F.R.S. (1687–1727): The Hanoverian man*, 1999.

[Coo06]   Eleanor Cook, *Enigmas and riddles in literature*, 2006.

[Cos00]   Shelly Costa, *"The Ladies' Diary." Gender and mathematics in England 1704–1754*, Ph.D. Thesis, Cornell University, 2000.

[Cos02a]   _____, *The ladies diary: Gender, mathematics and civil society in early eighteenth century england*, Osiris **17** (2002), 49–73.

[Cos02b]   Shelley Costa, *Marketing mathematics in early eighteenth-century England: Henry Beighton, certainty, and the public sphere*, Hist. Sci. **40** (2002), no. 2(128), 211–232, DOI 10.1177/007327530204000205. MR1958017

[Cot22]   Roger Cotes, *Opera miscellania*, 1722, Appendixed to Harmonia Mensurarum.

[Cow05]   Brian Cowan, *The social life of coffee: The emergence of the British coffeehouse*, 2005.

[Cri04]   Tony Crilly, *The Cambridge Mathematical Journal and its descendants: the linchpin of a research community in the early and mid-Victorian Age* (English, with English and French summaries), Historia Math. **31** (2004), no. 4, 455–497, DOI 10.1016/j.hm.2004.03.001. MR2105134

[Cro41]   T Crossley, *To diarians*, The Ladies' and Gentleman's Diary (1841), 18.

[Cru07]   A. D. D. Craik, *Mr. Hopkins' men*, Springer-Verlag London, Ltd., London, 2007. Cambridge reform and British mathematics in the 19th century. MR2327402

[Cur92]   Patrick Curry, *The naturalized female intellect*, Science in Context **5** (1992), 209–238.

[Cur07]   Louise Hill Curth, *English almanacs, astrology and popular medicine, 1550–1700*, Manchester University Press, 2007.

["D21]   "Dedascalus", *Female mathematicians*, 1821, pp. 27–35.

[Dal06]   Isaac Dalty, *A course of mathematics*, W. Glendinning, 1806.

[Dao95]   John Daotella, *Using riddles and interactive computer games to teach problem solving*, Teaching of Psychology **22** (1995), 33–36.

[Dar97]   Erasmus Darwin, *A plan for the conduct of female education in boarding schools*, 1797.

[Dav05]   William Davis, *The gentleman's mathematical companion*, 1805.

[Des02]   Sloan Evans Despeaux, *The development of a publication community: Nineteenth-century mathematics in British scientific journals*, ProQuest LLC, Ann Arbor, MI, 2002. Ph.D. Thesis, University of Virginia. MR2703853

[Des07]   Sloan Evans Despeaux, *Launching mathematical research without a formal mandate: the role of university-affiliated journals in Britain, 1837–1870* (English, with English and Italian summaries), Historia Math. **34** (2007), no. 1, 89–106, DOI 10.1016/j.hm.2006.02.005. MR2293232

[Des08]   Sloan Evans Despeaux, *Mathematics sent across the channel and the Atlantic: British mathematical contributions to European and American scientific journals, 1835–1900*, Ann. of Sci. **65** (2008), no. 1, 73–99, DOI 10.1080/00033790701663459. MR2417934

[Des14]   ———, *Mathematical questions: A convergence of mathematical practices in British journals of the eighteenth and nineteenth centuries*, Revue d'Histoire des Mathematiques **20(1)** (2014), 5–71.

[Dig56]   Leonard Digges, *A booke named tectonicon*, Felix Kingston, 1556.

[DM47]   Augustus De Morgan, *Arithmetical books from the invention of printing to the present time, being brief notices of a large number of works drawn up from actual inspection*, With an introduction by A. Rupert Hall, Hugh K. Elliott, Ltd., London, 1966. MR0202534

[Dra96]   Judith Drake, *An essay in defense of the female sex*, 1696.

[Dut93]   John Dutton, *Ladies' mercury*, 1693.

[Dut97]   ———, *Athenian mercury, London*, 1691–1697.

[E29]   Smith. D. E, *A source book in mathematics*, McGraw Hill, 1929.

[E58]   ———, *History of mathematics vol. II*, Dover Publications, 1958.

[EC17]   Nerida F. Ellerton and M. A. Clements, *Samuel Pepys, Isaac Newton, James Hodgson, and the beginnings of secondary school mathematics*, History of Mathematics Education, Springer, Cham, 2017. A history of the Royal Mathematical School within Christ's Hospital, London 1673–1868; With a foreword by Benjamin Wardhaugh. MR3642651

[Edw16]   Sue Bradford Edwards, *Hidden human computers: The black women of NASA*, ABDO Publishing, 2016.

[EG16]   Lindsey Eckert and Julia Grandison, *The almanac archive: Theorizing marginalia and duplicate copies in the digital realm*, Digital Humanities Quarterly **1:10** (2016), 1–16.

[Ell43]   Henry Ellis, *Original letters of eminent literary men of the sixteenth, seventeenth and eighteenth centuries*, The Camden Society, 1843.

[Eme43]   Robert Emerson, *The doctrine of fluxions: not only explaining the elements thereof, but also its applications*, J. Bettenham, 1743.

[Eme32]   ———, *English and British almanacs*, https://archive.org/details/fisheralmanacs, 1632–1832.

[Enr83]   Philip C. Enros, *The Analytical Society (1812–1813): precursor of the renewal of Cambridge mathematics* (English, with French and German summaries), Historia Math. **10** (1983), no. 1, 24–47, DOI 10.1016/0315-0860(83)90031-9. MR698136

[Eze92]   Margaret Ezell, *The gentleman's journal and the commercialization of restoration coterie literary practices*, Modern Philology **89** (1992), 323–341.

[Fea10]   Helen Forgasz et al., *International perspectives on gender and mathematics education*, Information Age Press, 2010.

[Fei84]   Mordechai Feingold, *The mathematicians' apprenticeship*, Cambridge University Press, Cambridge, 1984. Science, universities and society in England, 1560–1640. MR748485

[Fei90]   ———, *Mathematics and gender*, Teachers College Press, 1990.

[Fei05]   Timothy Feist, *The stationer's voice: The English almanac tradition in the early eighteenth century*, 2005.

[FRR00]   John Fauvel, *800 years of mathematical traditions*, Oxford Figures, Oxford Univ. Press, Oxford, 2000, pp. 1–27. MR1749701

[FRW11]   Raymond Flood, Adrian Rice, and Robin Wilson (eds.), *Mathematics in Victorian Britain*, Oxford University Press, Oxford, 2011. MR2885274

[Gas84]   John Gascoigne, *Mathematics and meritocracy: The emergence of the Cambridge mathematical tripos*, Social Studies of Science **14** (1984), 547–584.

[Gir80]   Mark Girouard, *Life in the English house*, Penguin Books, 1980.

[Gla80]   J.W.L. Glaisher, *Mathematical journals*, Nature **22** (1880), 73–75.

[Gol53]   Norman Goldsmith, *The Englishman's mathematics as seen in general periodicals in the eighteenth century*, Mathematics Teacher **46** (1953), 253–259.

[Goo78]   Esther Goody, *Questions and politeness*, Cambridge University Press, 1978.

[Gor02]   P. Gordon, *Geography anatomiz'd or the geographical grammar*, Robert Morden and Thomas Cockrill, 1702.

[Gor87]   Angeline Goreau, *Hers*, January 1987, p. 2.

[Gre17]   Matthew Green, *The surprising history of London's fascinating (but forgotten) coffeehouses*, March 2017.

[Gui89]   Niccolò Guicciardini, *The development of Newtonian calculus in Britain 1700–1800*, Cambridge University Press, Cambridge, 1989. MR1041793

[Gun63]   M. M. Gunning, *The histories of Lady Francis——s and Lady Caroline—— s*, J. Hoey and J. Pottss, 1763.

[Han51]   Nicholas Hans, *New trends in education in the eighteenth century*, Routledge and Kegan Paul, 1951.

[Har10]   Sylvia Harrop, *The enigma of Miss Nancy Mason of Clapham: Eighteenth century mathematician extraordinaire*, North Craven Heritage Trust Journal **19** (2010), 20–23.

[Hea52]   Robert Heath, *Truth triumphant or fluxions for the ladies*, W. Owen, 1752.

[Hea62]   ———, *The gentleman and lady's palladium*, H. Woodgate and S. Brooks, 1749–1762.

[Hec08]   Gregg Hecimovich, *Puzzling the reader: Riddles in nineteenth century British literature*, Peter Lang, 2008.

[Hei00]   J. L. Heilbron, *Geometry civilized*, The Clarendon Press, Oxford University Press, New York, 1998. History, culture, and technique. MR1638787

[Hey82]   T.W. Heyck, *The transformation of intellectual life in Victorian England*, Croom Helm, 1982.

[Hog76]   Edward R. Hogan, *George Baron and the "Mathematical Correspondent"* (English, with French summary), Historia Math. **3** (1976), 403–415, DOI 10.1016/0315-0860(76)90068-9. MR490768

[How82]   Albert Geoffrey Howson, *A history of mathematics education in England*, Cambridge University Press, Cambridge-New York, 1982. MR683878

[Huf07]   Cythnia Huffman, *Mathematical treasure: Francisco Algarotti's Newtonianism for the ladies*, January 2007, ejournal published by the Mathematical Association of America.

[Hut72]   Charles Hutton, *The principles of bridges*, T. Saint, 1772.

[Hut75]   ———, *The diarian miscellany*, G. Robinson and R. Baldwin, 1775.

[Hut89]   ———, *A supplement to The Ladies' Diary 1790–1791, a companion, or supplement to The Ladies' Diary: 1792–1801, the diary companion, being a supplement to The Ladies' Diary*, 1778–1789.

[Hut95]   ———, *A mathematical and philosophical dictionary. 2 vols*, J. Johnson, 1795.

[Hut98]   ———, *A course of mathematics. 2 vols*, J. Johnson, 1798.

[Hut12]   ———, *Tracts on mathematical and philosophical subjects. 3 vols*, F.C. and J. Rivington, 1812.

[Hut21]   ———, *On the mean density of the earth*, Transactions of the Royal Society **3** (1821), 276–292.

[Hut44]   _____, *Recreation in science and natural philosophy: Dr. Hutton's translation of Montucla's edition of ozanam*, 1844.

[Ita05]   Iona Italia, *The rise of literary journalism in the eighteenth century: Anxious employment*, Routledge, 2005.

[Jac01]   H. J. Jackson, *Marginalia: Readers writing in books*, Yale University Press, 2001.

[Jam11]   Kathryn James, *Reading numbers in early modern England*, Journal of the British Society for the History of Mathematics **26** (2011), 1–16.

[Joh89]   W. Johnson, *The Woolwich professors of mathematics, 1741–1900*, Journal of Mechanical Working Technology **18** (1989), 145–194.

[Jon15]   Clifford Jones, *The sea and the sky: The history of the Royal Mathematical School of Christ's Hospital*, Privately Published, 2015.

[Kea70]   Hugh Kearney, *Scholars and gentlemen: Universities and society in pre-industrial Britain 1500–1700*, Cornell University Press, 1970.

[Ken08]   Deborah Kent, *The Mathematical Miscellany and The Cambridge Miscellany of Mathematics: closely connected attempts to introduce research-level mathematics in America, 1836–1843* (English, with English and German summaries), Historia Math. **35** (2008), no. 2, 102–122, DOI 10.1016/j.hm.2007.12.001. MR2410156

[Kon12]   Ronnie Kon, *The Ladies' Diary, 18th century almanacs*, Word Ways **45** (2012), 248.

[Kro41]   David A. Kronick, *The Ladies' Diary 1704–1840*, 1704–1841.

[Kro91]   _____, *Scientific and technical periodicals of the seventeenth and eighteenth centuries: A guide*, The Scarecrow Press, 1991.

[Lan55]   John Landen, *Mathematical lucubrations containing new improvements in various branches of mathematics*, J. Nourse, 1755.

[Lan89]   Paul Landford, *A polite and commercial people: England 1737–1783*, Clarendon Press, 1989.

[Lan00]   _____, *Englishness identified: Manners and character 1650–1783*, Oxford University Press, 2000.

[Lap23]   P. S. Laplace, *Traité de 'ecanque céleste. 5 vols.*, 1798–1825.

[Led81]   Gilah Leder, *The Ladies' Diary*, Australian Mathematics Teacher **37(2)** (1981), 3–5.

[Led01]   _____, *Pathways in mathematics towards equity: a 25 year journey*, PMES **25** (2001), 1–43, ERIC 466733.

[Ley17]   Thomas Leybourn, *The mathematical questions proposed in The Ladies' Diary and their original answers, together with some new solutions, from its commencement in the year 1704 to 1816. 4 vols*, J. Mawan, 1817.

[Ley35]   _____, *The mathematical repository, new series. 5 vols*, 1804–1835.

[Loc93]   John Locke, *Some thoughts concerning education*, A. and J. Church, 1693.

[Lud85]   William Ludlam, *The rudiments of mathematics designed for students at the university*, John Evans, 1785.

["M19]   "Mathematicus", *Defense of English periodical works*, Philosophical Magazine **54** (1819), 367–369.

[MA21]   A. S. Hathaway, H. P. Manning, and R. C. Archibald, *Problems and Solutions: Problems: Solutions: 2801*, Amer. Math. Monthly **28** (1921), no. 6-7, 281–282, DOI 10.2307/2973352. MR1519814

[Mai00]   Brian Maidment, *Rearranging the year: The almanac, the day book and the year book and popular literary forms, 1789–1860*, St. Martin's Press, 2000.

[Mai13]   _____, *Beyond usefulness and ephemerality: The discursive almanac, 1828–60*, D. S. Brewer, 2013, British Literature and Print Culture.

[Mak73]   Bathsua Makin, *An essay to revive the antient education of gentlewomen*, J.D., 1673.

[Mar55]   Benjamin Martin, *The young gentleman's and ladies philosophys*, The General Magazine of Arts And Sciences **1** (1755), 1–7.

[Maz04]   Massimo Mazzotti, *Newton for ladies: gentility, gender and radical culture*, British J. Hist. Sci. **37** (2004), no. 2(133), 119–146, DOI 10.1017/S0007087404005400. MR2129460

[MBP82] Neil McKendrick, John Brewer, and J.H. Plumb, *The birth of a consumer society: The commercialization of eighteenth century England*, Indiana University Press, 1982.

[MG15] R. Montoito and A.V.M. Garnica, *Lewis Carroll, education and the teaching of geometry in Victorian England.*, History of Education **19** (2015), 9–27.

[Mie08] Anna Miegon, *The Ladies' Diary and the emergence of the almanac for women, 1704–1753*, Ph.D. Thesis, Department of English, Simon Fraser University, British Columbia, 2008.

[Moo81] Jonas Moore, *A new system of mathematicks*, A. Godbid, 1681.

[Mor16] Jennifer C. Mori, *Popular science in eighteenth century almanacs: The editorial career of Henry Andrews of Royston, 1780–1820*, History of Science **56** (2016), 19–44.

[Mye90] Sylvia Harcstark Myers, *The bluestocking circle: Women, friendship, and the life of the mind in eighteenth–century England*, Clarendon Press, 1990.

[New11] John Colton, ed. Newton, Isaac, *The young accomptant's debitor*, D. Brown, G. Straham, R. Simpson and T. Slater, 1711.

[New36] _____ , *The method of fluxions and infinite series*, Henry Woodfall, 1736.

[O'C99] Sheila O'Connell, *The popular print in England*, British Museum Press, 1999.

[Ott91] Charlotte Otten, *English women's voices, 1540–1700*, University of Florida Press, 1991.

[Ott17] Jessica Ottis, *"Set them to the cyphering schoole": Reading, writing, and arithmetical education, circa 1540–1700*, Journal of British Studies **56** (2017), 453–482.

[Pan87] M. Panteki, *William Wallace and the introduction of continental calculus to Britain: a letter to George Peacock* (English, with French and German summaries), Historia Math. **14** (1987), no. 2, 119–132, DOI 10.1016/0315-0860(87)90016-4. MR901868

[Pec94] J. Pechey, *The compleat herbal of physical plants*, Henry Bonwicke, 1694.

[Ped96] O. Pederson, *The "philomaths" of' 18th century England: A study in amateur science*, Centaurus **8** (1963), 238–262.

[Per77] Teri Perl, *The Ladies' Diary circa 1700*, Mathematics Teacher **70(4)** (1977), 354–358.

[Per79] Teri Perl, *The Ladies' Diary or Woman's Almanack, 1704–1841* (English, with French and German summaries), Historia Math. **6** (1979), no. 1, 36–53, DOI 10.1016/0315-0860(79)90103-4. MR518839

[Per86] Ruth Perry, *The celebrated Mary Astell*, University of Chicago Press, 1986.

[Phi90] Patricia Phillips, *The scientific lady. A social history of women's scientific interests 1520–1918*, Weidenfield and Nicolson, 1990.

[Pla08] John Playfair, *Traité de méchanique céleste* (Review), The Edinburg Review **11** (1808), 249–284.

[Plu76] N. Plumley, *The Royal Mathematical School within Christ's Hospital. The early years. Its aims and achievements*, Vistas Astronom. **20** (1976), no. 1-2, 51–59, DOI 10.1016/0083-6656(76)90009-X. MR490744

[Poo99] Steven Poole, *The riddle*, 1999, October 18, 1999.

[Pot12] Tiffany Potter, Ed., *Women, popular culture and the eighteenth century*, University of Toronto Press, 2012.

[PP91] S. Pumfrey and P.L. Rossi, *Science and popular belief in Renaissance Europe*, Manchester University Press, 1991.

[Rec43] Robert Recorde, *The ground of artes*, Reynold Wolff, 1543.

[Rey20] Myra Reynolds, *The learned lady in England 1650–1760*, Houghton Mifflin Co, 1920, Reprinted 1964.

[Ric88] Joan L. Richards, *Mathematical visions*, Academic Press, Inc., Boston, MA, 1988. The pursuit of geometry in Victorian England; With a foreword by I. B. Cohen. MR968441

[Ric96] Adrian Rice, *Mathematics in the metropolis: a survey of Victorian London* (English, with English, French and German summaries), Historia Math. **23** (1996), no. 4, 376–417, DOI 10.1006/hmat.1996.0039. MR1423694

[Ros75] Richard Ross, *The social and economic causes of the revolution in the mathematical sciences in mid–seventeenth century England*, Journal of British Studies **15** (1975), 46–66.

[Sar12]   Voula Sardikias, *For 'the present and future happiness of my dear pupils': The astronomical and educational legacy of Margaret Bryan*, Culture and Cosmos **16** (2012), 241–253.

[SC04]   William St. Clair, *The reading nation in the romantic period*, Cambridge University Press, 2004.

[Sch89]   Londa Schiebinger, *The mind has no sex? Women in the origins of modern science*, Harvard University Press, 1989.

[Sea01]   Jonathan Swift et al., *The works of the Rev. Jonathan Swift D.D.*, J. Johnson, 1801.

[Sha01]   Joanne Shallock, *Women and liberation in Britain 1800–1900*, Cambridge University Press, 2001.

[Sim37]   Thomas Simpson, *A new treatise of fluxions*, T. Gardner, 1737.

[SL79]   Frank Swetz and Hassan Lugglung, *Attitudes towards mathematics and their influencing factors*, International Education **9** (1979), 39–49.

[SLJ84]   Frank Swetz, Hassan Langgulung, and Abdul Rasid Johar, *Attitudes towards mathematics and school learning in Malaysia and Indonesia: Urban–rural and male–female dichotomies*, Comparative Education Review and reprinted: Akademia [Journal of Humanities and Social Sciences], 1983, 1984.

[Smi92]   Hilda Smith, *Intellectual base for feminist analysis. Women and reason*, University of Michigan Press, 1992.

[Smi15]   Michael Smith, *The debs of Bletchley Park*, Aurum Press, 2015.

[Smy08]   Adam Smyth, *Almanacs, annotators, and life-writing in early modern England*, English Literary Renaissance **38** (2008), 200–244.

[Sob95]   Dava Sobel, *Longitude! The true story of a lone genius who solved the greatest scientific problem of his time*, Walker and Company, 1995.

[Sta60]   Magdalen Stanfford, *Home–life of English ladies in the seventeenth century*, Bell and Doldy, 1860.

[Ste99]   Larry Stewart, *Other centers of calculation or where the royal society didn't count: commerce, coffee–houses and natural philosophy in early modern England*, British Journal for the History of Science **32** (1999), 133–153.

[Sto12]   Lawrence Stone, *Mathematical expeditions: Exploring word problems across the ages*, Harper and Row, 2012.

[Swe89]   Frank Swetz, *Cross–cultural insights into the question of male superiority in mathematics: Some Malaysian findings*, Mathematics, Education and Society [Science and Technology Education] Paris: UNESCO **35** (1989), 139–140.

[Swe90]   ———, *Psychology and socio–economic development*, National University of Malaysia Press, 1990.

[Swe08]   Frank J. Swetz, *The mystery of Robert Adrain*, Math. Mag. **81** (2008), no. 5, 332–344, DOI 10.1080/0025570x.2008.11953574. MR2473870

[Swe12]   Frank J. Swetz, *Mathematical expeditions*, Johns Hopkins University Press, Baltimore, MD, 2012. Exploring word problems across the ages. MR2952821

[Tat69]   Walter Tattner, *God and expansionism in Elizabethan England: John Dee, 1527–1583*, Journal of the History of Ideas **25** (1969), 17–34.

[Tay51]   Archer Taylor, *English riddles from oral tradition*, University of California Press, 1951.

[Tay54]   E. G. R. Taylor, *The mathematical practitioners of Tudor & Stuart England*, Cambridge, at the University Press, 1954. MR0066999

[Tay66]   ———, *The mathematical practitioners of Hanoverian England, 1714–1840*, Cambridge University Press, 1966.

[Tha43]   Arnold Thacker, *A miscellany of mathematical problems, vol. I*, Thomas Aris, 1743.

[Tha08]   Arnold Thackray, *"John Dalton." Complete dictionary of scientific biography*, Charles Scribner's Sons, 2008.

[Tom71]   Richard S. Tompson, *Classics or charity: The dilemma of the 18th century grammar school*, Manchester University Press, 1971.

[Top 5] John Toplis, *On the decline of mathematical studies*, Philosophical Magazine **20** (1804–5), 23–31.

[Tur45] John Turner (ed.), *Mathematician*, 1745.

[Tur52] John Turner, *Mathematical exercises*, 1750–1752.

[Upt13] Chris Upton, *The birthplace of the 'diversion of the fair sex'*, Birmingham Post (2013), 66, Lifestyle section.

[Vn50] A. Vickery n.d. (ed.), *Women advising women. Advice books, manuals and journals for women, 1450–1837*, Adam Matthew Publications, 1450, Microfilm. Part 2: reels.

[Wal73] Obadiah Walker, *Of education especially for young gentlemen*, 1673.

[Wal73] P. J. Wallis, *British Philomaths—mid-eighteenth century and earlier*, Centaurus **17** (1973), no. 4, 301–314, DOI 10.1111/j.1600-0498.1973.tb00201.x. MR446831

[Wal86] R.V. Wallis, *Biobibliography of British mathematics and its applications*, Epsilon Press, 1986.

[Wal97] Alice Walters, *Conservation pieces: Science and politeness in eighteenth century England*, History of Science **35** (1997), 121–154.

[War83] Leland Warren, *Turning reality round*, Eighteenth–Century Life **8** (1983), 65–87.

[War09] Benjamin Wardhaugh, *Mathematics in English printed books, 1473–1800: a bibliometric analysis*, Notes and Records Roy. Soc. London **63** (2009), no. 4, 325–338, DOI 10.1098/rsnr.2008.0033. With supplementary data available online. MR2558893

[War12] _____ , *Learning geometry in Georgian England*, Convergence Ejournal of Mathematical Association of America (2012), August, 2012.

[War17] Benjamin Wardhaugh, *Charles Hutton: 'one of the greatest mathematicians in Europe'?*, BSHM Bull. **32** (2017), no. 1, 91–99, DOI 10.1080/17498430.2016.1236319. MR3610361

[War18] _____ , *The correspondence of Charles Hutton (1737–1823): Mathematical networks in Georgian Britain*, Oxford University Press, 2018.

[Wer17] Jacqueline D. Wernimont, *Poetica–mathematical women and the learning disabled*, pp. 337–350, Palgrave, Macmillan, 2017.

[Whi14] William Whiston, *A new method for discovering the longitude both at sea and land, humbly proposed to the consideration of the public*, J. Johnson, 1714, M. A. sometime Professor of the Mathematicks in the University of Cambridg. and Humphry Ditton, Master of the New Mathematick School in Christ's Hospital, London.

[Whi70] Cynthia White, *Women's magazines 1693–1968*, Michael Joseph, 1970.

[Whi18] Mathew White, *Discovering literature: Restoration and the 18th century*, British Library [website], 2018, 21 June.

[Wil48] John Wilkins, *Mathematical magick*, E. Gellibrand, 1648.

[Wil48] Thomas Wilkinson, *Mathematical periodicals*, Mechanics Magazine **48** (1848), 56–57, 83–84, 154–155, 224–226, 254–255, 279–281, 342–343, 401–402, 466–468, 514, 583.

[Wil03] A.N Wilson, *The Victorians*, W.W. Norton, 2003.

[Wil14] Lisa Wilde, *Whiche ells shuld faire exedle mans mynd: Numerical reasoning in Robert Records ground of arts (1543)*, Journal of the Northern Renaissance. (ejournal) (2014).

[Wol87] Mary Wollstonecraft, *Thoughts on the education of a daughter*, Joseph Johnson, 1787.

[Wol92] _____ , *A vindication of the rights of women*, Peter Edes, 1792.

[Woo42] John Woodhouse, *A new almanack and prognostication for the year of our lord 1642*, Dawson, 1642.

[Woo41] Wesley Woolhouse, *The lady's and gentleman's diary*, The Company of Stationers, 1841.

[Wor77] Wordsworth, *Scholae academicae: Some account of studies at the English universities in the eighteenth century*, Cambridge University Press, 1877.

[WP80] Ruth Wallis and Peter Wallis, *Female philomaths* (English, with French and German summaries), Historia Math. **7** (1980), no. 1, 57–64, DOI 10.1016/0315-0860(80)90064-6. MR559836

[Zwe11] Karen Zwer, *John Dalton's puzzles: From meteorology to chemistry*, Studies in the History and Philosophy of Science **42** (2011), 58–66.

Endnotes

Foreword.

1 Joan W. Scott, "Gender: A Useful Category of Historical Analysis," American Historical Review 91 (1986): 1053–1075, http://www.jstor.org/stable/1864376.

Preface.

1 See for example, Fenneman, 1990.

2 My surveys among female students in Southeast Asia would seem to affirm this theory. See, for example Swetz and Luggling, 1979. [**SL79**]

3 Perl, 1977. [**Per77**]

4 Perl, 1979. [**Per79**]

5 Costa, 2000.

6 Albree and Brown, 2009.

7 Heilon, 2000.

8 Swetz, 2012, Chapter 12: *The Ladies' Diary* (1704-1841). pp. 115-122. [**Swe12**]

Chapter 1. A Fortuitous Encounter.

1 The coffee house served as a men's club. Besides its social function, the coffee house was a center for political, commercial, and educational activities. At the time the *Diary* existed, it was estimated that London contained about 3000 coffee houses. The phenomenon of the coffee house is discussed in Green, 2017 [**Gre17**]. See also White, 2018. [**Whi18**]

2 For a delightful history of coffee and its impact, see Cowan, 2005. [**Cow05**]

3 *The Ladies' Diary*, [LD] 1754, p. 28.

4 LD, 1755, p. 46.

5 Martin, 1755, p. 2. [**Mar55**]

6 In this period of history, written conventions for the English language are still in a state of flux and development. The punctuation within the title of this journal will change several times during its lifetime. Throughout this work, for convenience, I will refer to the periodical as: *"The Ladies' Diary,"* its original name, *"Diary"* or by the English reader's nickname *"Dia."*

Chapter 2. *The Ladies' Diary: or Woman's Almanack:* A New Kind of Publication

For those unfamiliar with this publication, just what was *The Ladies' Diary*?

1 See Italia, 2005. [**Ita05**]

2 Perkins, 1996, p.14.

3 Miegon, op cit., p.93.

4 LD, 1718, p.3.

5 LD, 1726, pp. 1-2.

6 Ellis, 1843, p.304.

7 Thomas Kirkman is best remembered for posing "The Schoolgirl Problem"—Fifteen young ladies in a school walk out three abreast for seven days in succession: it is required to arrange them daily, so that no two shall walk twice abreast.

8 His comments are quoted in Archibald, 1929, pp.379–80.

9 Playfair, 1808.

Chapter 3. *The Ladies' Diary*: Conception and Evolution

How did the editors shape and control the direction in which the *Diary* developed and progressed?

1 For a detailed discussion, see Capp, 1979. [**Cap79**]

2 See Perkins, 1996.

3 Woodhouse, 1642. [**Woo42**]

4 LD, 1705, pp. 24–25.

5 This theory is proposed in Miegon, 2008, p.100. [**Mie08**]

6 See correspondence as given in *Original Letters of Eminent Men* …, 2016, pp.304–315. [**Ell43**]

7 As referenced in LD, 1709, p.11. A more concise definition is supplied by Cook, 2006, p.260 [**Coo06**]: "A short composition in prose or verse, in which something is described by intentionally obscure metaphor, in order to afford an exercise for the ingenuity of the reader in giving what is meant …"

8 *Original letters*, op.cit., p.309.

9 See Landford, 2000 [**Lan00**], for further discussion.

10 LD, 1709, p.15.

11 LD, 1707, p.37.

12 LD, 1709, p.3.

13 LD, 1710, p. 36.

14 LD, 1711, p.31. The prize was ten or twelve free copies of *The Ladies' Diary*.

15 A *Delight for the Ingenious*, a collection of poems was published in London in 1648. The complier merely lists himself as "R.B.", author of *A History of the Wars of England*. No record of this book can be found.

16 For a more detailed discussion of the life and works of Henry Beighton, see Cook, 1999. [**Coo99**]

17 LD, 1715, p.12.

18 LD, 1718, pp.1-2. ("Introduction to the Second Part").

19 The engraving was entitled "The Engine for Raising Water with a Power made by Fire." This was the first published illustration of this new, revolutionary machine. Beighton's picture became a model for other diagrams of steam engines.

20 The first engine operated at 12 strokes/min., generating 5.5 horsepower and raised 10 gallons of water from a depth of 150ft., per stroke. Beighton published a table listing the quantity of water raised by an engine with a six- foot stroke working at 16 strokes/min. LD, 1721, pp.21–23.

21 LD, 1721, p.21.

22 These concerned: an analysis of the waterworks at London Bridge (1731); a description of a plane table he devised and some meterological observations (1735–1736) and an easy, cheap way of constructing a large quadrant (1739).

23 LD, 1720, Preface.

24 For further discussion, see Bataille, 1990.

25 See Bataille, 1987.

26 Ibid. See also Cajori 1919, pp.219–222. [**Caj19**]

27 LD, 1748, p.21.

28 *Mechanics Magazine, Museum, Register, Journal and Gazette* (1848) 50:267.

29 LD, 1746, pp.40-45.

30 LD, 1751, pp.43–45; 1752, pp.44–45. For a further analysis of these entries see Cajori, 1919, pp.219–222. [**Caj19**]

31 LD, 1752, problems 359, 360, pp.25–26.

32 *Mechanics Magazine*, op.cit. p.270.

33 LD, 1749, quoted Ibid. p.467.

34 *The Monthly Review or New Literary Journal*, December, 1750, pp.129–131.

35 Cotes, 1722, pp.1–22 [**Cot22**]. This charge seemed to be unfounded and concerned "the fluxions of the sides and angles of a spherical triangle."

36 See discussion in *Mechanics Magazine* (1849), 50:9.

37 A detailed discussion of the controversy can be found in Cajori, op.cit., "Robert Heath and Friends of Emerson in Controversy with John Turner and Friends of Simpson," Chapter XIII, pp.207–223.

38 *The Ladies' Chronologer*, 1754.

39 De Morgan, 1847, pp.71–72.

40 Thomas Wilkinson provides a tirade against Robert Heath and his editorship of the *Diary* in the *Mechanic Magazine and Journal of Science, Arts and Manufacture*, 1849, 50:468–475. The particular rhyme is found on p.47 [**Wil48**]

41 An excellent biography of Robert Simpson can be found in: *The Practical Mechanic and Engineers' Magazine*, 1842, 16:295–296.

42 A riddle or puzzle made up of letters, pictures or symbols whose names sound like the parts or syllables of a word or phrase which is to be discovered.

43 The first published *Tract of the Royal Society*.

44 See Wardhaugh, 2017. [**War17**]

45 Charade: a form of literary riddle popularized in nineteenth-century France where each syllable of the answer is given as a separate word.

46 Hutton had been most favorably impressed by a manuscript written by the young Gregory on the "Use of the Sliding Rule."

47 For example, page 49 of the 1835 edition contained "Horae Geometricae", an article by T.S. Davis F.R.S. on spherical geometry. Thomas Stephen Davis was a well-respected mathematician and a

Mathematics Master at RMA Woolwich. The 1837 edition contained "On Summation of Series" by a Mr. Rutherford, identified only as being from Berwick upon Tweed; however, Rutherford also taught at Woolwich and became known in 1841 for his 208-digit estimation of the mathematical constant π. Also included in this issue were two reprinted papers that had been published by the Royal Society.

Chapter 4. "Delightful and Entertaining Particulars"—Problem Solving

Why enigmas and mathematical problems?

1 Gregg Hecimovich, "Puzzling the Reader: Riddles in Nineteenth-Century British Literature." Studies in19th Century British Literature, 2008, vol. 26, 136, p.6 [**Hec08**]

2 Ibid. p.14.

3 LD, 1835, p.27, p.30.

4 Answer, A pillow.

5 Hadley and Singmaster, 1992.

6 Ibid. p.111.

7 They arrive: Martin, John, Stephen, Geoffrey.

8 For example, Emma, 1815, vol. I, Chapter 9.

9 Answer: A frying pan.

10 Leybourn, vol. I, p.1.

11 LD, 1709, p.14.

12 Leybourn, vol. I, p.1.

13 LD, 1710, p.33.

14 Leybourn, vol. 2, p.69.

15 This practice seems to have been followed in various periodicals of this time that actively solicited and published correspondence.

16 Leybourn, vol.2, p.88.

17 Ibid. p.105.

18 Ibid. p.154.

19 Leybourn, op.cit. vol. I. p.2, p.5, p.3, respectively.

20 Leybourn, Ibid., p.88, p.40, p.151, p.167, respectively.

21 Hutton, Darian Miscellany, vol. I, p.117.

22 LD, 1716, p.27.

23 Leybourn, op cit. vol. I, p.414; vol. II, p.6, p.15, p.102, respectively.

24 Manning and Archibald, 1921, pp.281-282. [**MA21**]

25 Leybourn, Ibid. Vol. II p.151, p.175, p.180, respectively.

26 Ibid. p.221, p.240, p.246, p.298, respectively.

27 Ibid. p.399, vol. III, p250, vol. II, p.284, vol. III, p.397, vol. III, p.400.

28 LD, 1835: p.47, problems 1590, 1591, p.48, problem 1597.

Chapter 5. Mathematics, Education, and Women in Eighteenth- and Nineteenth-Century England

What were women's opportunities to study and know mathematics?

1 The French king Francis I had established Royal chairs of astronomy and geometry in order to support navigation. Elizabeth was advised to do the same. See Ross, 1975, p.48. [**Ros75**]

2 See Wilde, 2014. [**Wil14**]

3 Recorde, 1543, p.33. In an excerpt from this dialogue, we find the Master responding to his young charge's inquiry by probing the youth as to the value of arithmetic:

Master: Wherefore in all great works are Clerks so much desired? Wherefore are Auditors so well fed? What causeth Geometricians So highly to be enhaunsed? Why are Astronomers so greatly advanced? Because that by Number such things they finde, which else would farre excel mans minde. Scholer: Verily, sir, if it bee so, that these men by numbering, Their cunning do attain, at whose great works most men do wonder, then I see well I was deceived, and numbering is a more cunning thing than I took it to be.

4 Leonard Digges served his patron, Robert Dudley, Earl of Leichester, advising on nautical affairs and military engineering.

5 Henry Billingsley was a wealthy London haberdasher and had served as Mayor of London.

6 Billingsley, Preface.

7 Wardhaugh, 2009.

8 Cassels, 1979. [**Cas80**]

9 See Taylor, 1966 [**Tay66**]

10 See Plumley, 1976 [**Plu76**]; Ellerton and Clements, 2017. [**EC17**]

11 Discussed in detail in Allen, 1970.

12 Jones, 1851, p1.

13 See Johnson,1989 [**Joh89**]

14 As quoted in Ball, 1889, pp.41–42. [**Bab30**]

15 Taylor, 1966, op. cit. [**Tay66**]

16 Taylor, 1966, op. cit. [**Tay66**]

17 For a discussion see Guicciardini, ibid. pp.135–138. [**Gui89**]

18 *The Connoisseur*, 1774, p.19; quoted in Gascoigne, 1984, p.566 [**Gas84**]. The Saunderson referred to in the note is Nicholas Saunderson (1682–1739) a blind English scientist and mathematician. He held the position of Lucasian Professor at Cambridge University.

19 *Home-Life of English Ladies in the Seventeenth Century*, 1860, pp.78–79. The author is merely identified as also being the author of "Magdalen Stafford." Further research affirms that the mysterious author actually is Magdalen Stafford.

20 For more of Mary's work see Mary Serjant, "Mathematical Treasures", *Convergence*.

21 *Arithmetic Exercise Book*, Folger Shakespeare Library ref. vb 292.

22 As quoted in Otten, 1991, p.4.

23 Hans, 1951, p.202.

24 Ibid. p.203.

25 As quoted by Saridkias, 2011, p.241.

26 "Didascalus", 1821, p.33.

27 "Bipliopola", 1822, p.119.

28 Ibid., p.120.

29 *Princess Ida* opened in London on January 5, 1884. The opera satirized women's education and feminism.

Chapter 6. *The Ladies' Diary* as a Facet in the Mathematical and Scientific Transition of the Era

What intellectual value did the *Diary* possess?

1 *York Courant*, Oct., 1828. As quoted in Black and Howson, 1979, p.91. [**BH79**]

2 *York Courant*, 1830. As quoted in Black and Howson, 1979, p.92. [**BH79**]

3 Leybourn,1817, reprinted, in part, some of Hutton's previous corrections to solutions: they will be indicated by a concluding "H." Leybourn's own corrections are marked with an "L"; all other corrections bear the names of their original *Diary* correspondents.

4 Leybourn, 1817, vol I, p.45.

5 Ibid. p.227.

6 Ibid. p.246.

7 Leybourn, 1817, op. cit. vol III, p.99.

8 *The Ladies' Diary*, 1826, p.47.

9 LD, 1710. p.4.

10 LD, 1706, 34-36.

11 LD, 1709, pp.3-4.

12 Respectively: LD, 1772, p.31, LD, 1786, p.32; LD, 1807, p.32; LD,1825, pp.32.

13 LD, 1826, p.32.

14 LD, 1709, pp.21-23.

15 Cajori, 1919, pp.219-222.

16 LD, 1838, pp.49-72.

17 LD, 1837, pp.60-69.

Chapter 7. *"Dia"* as a Mathematical Testament

Did the *Diary* reflect and support the mathematical reforms taking place during the span of its publication?

1 Isaac Newton produced a technique of differentiation openly modeled on the concept of change. In his *Method of Fluxions* written in 1671, he considers a curve as the path of a moving point, thus the coordinates of a point are in a constant state of change. Newton termed a changing quantity a *fluent* and its rate of change the *fluxion* of the fluent. If a fluent is given by y, then in modern terms, the fluxion would be represented by dy/dt.

2 George Berkeley was an outspoken critic of Isaac Newton and his calculus. His most well-known attack was published in 1734 and entitled The Analyst. [**Ber34**]

3 Colson, 1801, p.ii.

4 Swift's words were: "The longitude: Missed on by wicked Will Whiston, And not better hit on by good Master Ditton. So Ditton and Whiston may both be pi-ed on, And Whiston and Ditton may

both be sh-t on … " The chorus repeats/echoes the insult. Swift, Sheridan and Nichols, 1801, XVII:429. [**Sea01**]

5 For a discussion of this saga, see Sobel, 2007.

6 Wikisource, "The Ballad of Grisham College", 26.

7 See Richards, 1988. [**Ric88**]

8 Guicciardini, op. cit., p.101.

9 Toplis, 1804-1805, p.28. [**Top 5**]

10 Playfair, 1808, p.281. [**Pla08**]

11 See discussion given in Guicciardini, op.cit., pp.135-138.

12 Playfair, 1808, p.282.

13 Leybourn [Ley.], 1817, vol. I, p.36.

14 Ley. Ibid., p.284.

15 Ley. 1817, vol. II, pp.178-180.

16 Ley. 1817, vol. III, pp.11-13.

17 Ley. 1817,vol. IV, pp.11-12.

18 LD, 1827, p.48.

19 Ley. 1817,vol. I, pp.5-8.

20 Ley. 1817,vol. II, pp.24-26.

21 Ley. 1817, vol. IV, pp.1-3.

22 LD, 1825, p.47.

23 Ley, 1817, vol. IV, pp.228-231.

24 LD, 1833, p.48.

25 LD, 1834, p.45.

26 LD, 1834, p.35.

27 Ibid. p.47.

28 LD, 1838, p.47.

29 Ley. 1817, p.vii.

30 Ley. 1817, vol. II, pp.93-94.

31 Ley. 1817, vol. IV, pp.70-71.

32 Ibid. pp.176-177.

33 Ley. 1817, vol. III, pp.335-336.

34 "*Mathematicus*", 1819, pp. 367–371. ["**M19**]

35 Ibid.

36 Henry Meikle was a frequent contributor to the philosophical magazine. Little further is known about him. Most of his contributions concerned physics, particularly that of heat and its transfer. His discussions were often a center of controversy and challenged by other scientists, for example, James Ivory, FRS (1765–1842).

37 "*Mathematicus*", 1819, op.cit. p.368

38 Clarke, 1877, p.70. Mr. Clarke's mathematical accomplishments were modest as compared to some other *Diary* correspondents. One man who established his interest and talent by correctly answering over one hundred mathematical quires in the *Diary* and other various publications was the Rev. Charles Wildbore (1737-1802). Rev. Wildbore, described as "an ingenious mathematician," at times wrote under the names "Eumenes" and "Amicus" served as vicar for a parish in Broughton Sulney besides running an academy for young men. His mathematical talents were

admired by Charles Hutton who, in 1780, secured the clergaman the editorship of *The Gentle-man's Diary*. Rev. Mr. Wildbore was offered membership in The Royal Scientific Society but he declined, preferring to remain "a humble village pastor." Chalmer's Biography, 1812. Vol. 32:54. [**Cha12**]

39 Ley, 1817, Vol. I: vii.

40 Guicciardini, op.cit., p.116.

Chapter 8. Women and the *Diary*

Did the *Diary* really serve the needs of women?

1 Ellis,1843. A personal correspondence with Humphrey Wanley: November 8, 1703, pp.304-315.

2 Humphrey Wanley F.R.S. (1672-1726). A London-based librarian, scholar of the English lan-guage and recognized philosopher.

3 Ellis, op.cit. p.307.

4 Leybourn, 1817, p.viii.

5 Reynolds, 1964. Op.cit. p.328. [**Rey20**]

6 Bataille, 1987. "Sophia Western" was the heroine in Henry Fielding's novel *Tom Jones* (1749).

7 Costa, 2002, pp.189-201.

8 Ibid. p.189.

9 Harrop, 2010, p.21. [**Har10**]

10 LD, 1758, p.46.

11 Although they both signed themselves as "Mrs.," Tipper identifies them as sisters. LD, 1711, p.39. Perhaps such identification was a social protection intended to dissuade male suitors.

12 Harrop, 2010, p.23.

13 LD, 1796, p.22 ans.: 6.32455, 9.48683, 12.24745, 14.83239.

14 For example, the "Tinker Problem", LD,1711, p.33, Barbara Sidway proposed an extension of the problem, LD, 1712, p.33.

15 LD, 1715, p.40

16 LD, 1716, p.31.

17 See "Bibliapola", 1822. ["**B22**]

18 Perl, 2008, op.cit. pp.28-30; Costa, 2000, op.cit. p.150 and Meigon, 2008, op.cit. p.291.

19 Italia, 2005, p.10. [**Ita05**]

20 There was a prevailing belief that women, "being the daughters of Eve," were intrinsically morally weak.

21 Goreau, 1987.

22 Mary Somerville was Great Britain's first notable woman scientist. She was the second female to have a paper published in the Tracts of the Royal Society, 1826; the astronomer Caroline Herschel preceded her in 1787. For a more complete examination of her accomplishments, see, Bruck, 1996. [**Bru96**]

23 Miss Fawcett was even immortalized in an anonymous poem of 1890. The first two stanzas of which are: Hail the triumph of the corset Hail the fair Philippa Fawcett Victress in the fray Crown her queen of Hydrostatics And the other Mathematics Wreathe her brow with bay.
If you entertain objections To such things as conic sections Put them out of sight Rather sing of the essential Beauty of the Differential Calculus tonight.

Chapter 9. *The Ladies' Diary*, a Noteworthy Heritage

So what did we learn about the *Diary*'s effects and societal impact?

1 "Mathematicus", 1891, p.368.

2 As quoted by John Baily in his "Memoir", 1877, p.cxv.

3 Miegon. 2008, pp.91-92.

4 Glaisher, 1880, p.74. [**Gla80**]

5 Glaisher, 1880, p.74.

6 See Swetz, 2008. [**Swe08**]

7 For a complete discussion of Gill's attempts to establish American journals, see Kent, 2008, pp. 106-108. [**Ken08**]

8 Thackray, 2008.

9 Zwer, 2011. [**Zwe11**]

10 Ley. 1817, I:v.

11 See Smith, 2015. [**Smi15**]

12 As discussed in Edwards, 2016. [**Edw16**]